INTERNATIONAL AS AND A LEVEL
PHYSICS

Richard Woodside
Edited by Mary Jones and Chris Mee

Hachette UK's policy is to use papers that are natural, renewable and recyclable products and made from wood grown in sustainable forests. The logging and manufacturing processes are expected to conform to the environmental regulations of the country of origin.

Orders: please contact Bookpoint Ltd, 130 Milton Park, Abingdon, Oxon OX14 4SB. tel: (44) 01235 827827; fax: (44) 01235 400401; email: education@bookpoint.co.uk. Lines are open 9.00–5.00, Monday to Saturday, with a 24-hour message answering service. Visit our website at www.hoddereducation.co.uk

© Richard Woodside 2011
First published in 2011 by
Hodder Education, a Hachette UK company
338 Euston Road
London NW1 3BH

Impression number   5 4 3 2 1
Year   2014 2013 2012 2011

All rights reserved. Apart from any use permitted under UK copyright law, no part of this publication may be reproduced or transmitted in any form or by any means, electronic or mechanical, including photocopying and recording, or held within any information storage and retrieval system, without permission in writing from the publisher or under licence from the Copyright Licensing Agency Limited. Further details of such licences (for reprographic reproduction) may be obtained from the Copyright Licensing Agency Limited, Saffron House, 6–10 Kirby Street, London EC1N 8TS.

Illustrations by Greenhill Wood Studios
Typeset in ITC Leawood 8.25 pt by Greenhill Wood Studios
Printed by MPG Books, Bodmin

A catalogue record for this title is available from the British Library

ISBN 978 1 4441 1269 6

International AS and A Level Physics Revision Guide

Contents

Introduction
About this guide 5
The syllabus 5
Assessment 6
Scientific language 8
Revision 9
The examination 10

AS Content Guidance
I General physics 14
II Newtonian mechanics 24
III Matter 54
IV Oscillations and waves 65
V Electricity and magnetism 80
VI Modern physics 96
AS Experimental Skills and Investigations 100

AS Questions & Answers
About this section 118
Exemplar paper 119

A2 Content Guidance
I General physics 132
II Newtonian mechanics 132
III Matter 144
IV Oscillations and waves 160
V Electricity and magnetism 170

VI Modern physics 201
VII Gathering and communicating information 221
A2 Experimental Skills and Investigations.. 258

■ ■ ■

A2 Questions & Answers
Exemplar paper: section A... 268
Exemplar paper: section B... 282

Introduction

About this guide

This book is intended to help you to prepare for your University of Cambridge International A and AS level physics examinations. It is a revision guide, which you can use alongside your textbook as you work through your course and towards the end when you are revising for your examination.

The guide is split into two main parts. Pages 13 to 130 cover the AS examination. Pages 131 to 287 cover the A2 examination.

- This **Introduction** contains an overview of the AS and A2 physics courses and how they are assessed, some advice on revision and advice on taking the examinations.
- The **Content Guidance** sections provide a summary of the facts and concepts that you need to know for the AS or A2 physics examination.
- The **Experimental Skills** sections explain the data-handling skills you will need to answer some of the questions in the written examinations. It also explains the practical skills that you will need in order to do well in the practical examination.
- The **Questions and Answers** sections contain a specimen examination paper for you to try. There are also two sets of students' answers for each question, with typical examiner comments.

It is entirely up to you how you use this book. We suggest you start by reading through this Introduction, which will give you some suggestions about how you can improve your knowledge and skills in physics and about some good ways of revising. It also gives you some pointers into how to do well in the examination. The Content Guidance will be especially useful when you are revising, as will the Questions and Answers.

The syllabus

It is a good idea to have your own copy of the University of Cambridge International Examinations (CIE) A and AS level physics syllabus. You can download it from:

http://www.cie.org.uk

The **Syllabus Content** provides details of the facts and concepts that you need to know, so it is worth keeping a check on this as you work through your course. The AS syllabus is divided into six sections, I to VI; the A2 syllabus is divided into seven sections, I to VII. Each section contains many learning outcomes. If you feel that you have not covered a particular learning outcome, or if you feel that you do not understand it, it is a good idea to do something to correct this at an early stage. Don't wait until revision time!

Introduction

Do look through all the other sections of the syllabus as well. There is a useful summary of the quantities you should be familiar with and their usual symbols and the unit in which they are measured. As you work through your course, you could use a highlighter to mark each of these quantities that are relevant for the topics you have covered.

Syllabus content

In the syllabus the AS work is written in ordinary type, whereas the A2 work is written in **bold** type.

The content of the AS syllabus is divided into six sections:
I   **General physics** — Physical quantities and units; Measurement techniques
II **Newtonian mechanics** — Kinematics; Dynamics; Forces; Work, energy and power
III **Matter** — Phases of matter; Deformation of solids
IV **Oscillations and waves** — Waves; Superposition
V   **Electricity and magnetism** — Introductory electric fields; Current electricity; d.c. circuits
VI **Modern physics** — Nuclear physics

The content of the A2 syllabus is divided into seven sections:
I   **General physics** — Physical quantities and units; Measurement techniques
II **Newtonian mechanics** — Motion in a circle; Gravitational field
III **Matter** — Ideal gases; Temperature; Thermal properties of materials
IV **Oscillations and waves** — Oscillations
V   **Electricity and magnetism** — Electric fields; Capacitance; Magnetic fields; Electromagnetism; Electromagnetic induction; Alternating currents
VI **Modern physics** — Charged particles; Quantum physics; Nuclear physics
VII **Gathering and communicating information** — Direct sensing; Remote sensing; Communicating information

The main part of this book, the Content Guidance, summarises the facts and concepts covered by the learning outcomes in all of these 13 sections.

Assessment

The AS examination can be taken at the end of the first year of your course, or with the A2 examination papers at the end of the second year of your course.

What is assessed?

Both the AS and A2 examinations will test three Assessment Objectives. These are:

A: Knowledge with understanding
This involves your knowledge and understanding of the facts and concepts described in the learning outcomes in all sections. Questions testing this Assessment Objective will make up 37% of the whole examination.

B: Handling information and solving problems
This requires you to use your knowledge and understanding to answer questions involving unfamiliar contexts or data. The examiners ensure that questions testing this Assessment Objective cannot have been practised by candidates. You will have to *think* to answer these questions, not just remember! An important part of your preparation for the examination will be to gain confidence in answering this kind of question. Questions testing this Assessment Objective will make up 40% of the whole examination.

C: Experimental skills and investigations
This involves your ability to do practical work. The examiners set questions that require you to carry out experiments. It is most important that you take every opportunity to improve your practical skills as you work through your course. Your teacher should give you plenty of opportunity to do practical work in a laboratory.

The skills built up in AS are developed further at A2. In addition, you are expected to understand how to plan an investigation. Although Paper 5 does not require you to carry out an experiment, the only way to learn the skills required to succeed on this paper is by working hard in the laboratory. Questions testing this Assessment Objective will make up 23% of the whole examination.

Notice that more than half the marks in the examination — 63% — are awarded for Assessment Objectives B and C. You need to work hard on developing these skills, as well as learning facts and concepts. There is guidance about Assessment Objective C for AS on pages 100–116, and for A2 on pages 258–265.

The examination papers

For security reasons there are now papers for different time zones across the world. The papers are labelled 11, 12 etc., 21, 22 etc., 31, 32 etc. For convenience this book will use Paper 1 for the suite of papers 11, 12 etc.; Paper 2 for the suite 21, 22 etc. and so on.

The AS examination has three papers:
- Paper 1   Multiple choice
- Paper 2   Structured questions
- Paper 3   Advanced Practical Skills

Paper 1 and Paper 2 test Assessment Objectives A and B. Paper 3 tests Assessment Objective C.

Paper 1 contains 40 multiple-choice questions. You have 1 hour to answer this paper. This works out at about one question per minute, with time left over to go back through some of the questions again.

Paper 2 contains structured questions. You write your answers on lines provided in the question paper. In numerical questions, you are given a blank area for your calculation, with an answer cue at the end. The answer cue will remind you of the quantity you are to calculate, followed by a short line for the numerical answer. The unit may or may not be given. Watch out for this. If no unit is given, you must provide it. You have 1 hour to answer this paper.

Paper 3 is a practical examination. You will work in a laboratory. As with Paper 2, you write your answers on lines provided in the question paper or in the blank areas provided for tables and numerical work. You have 2 hours to answer this paper.

The A2 examination has two papers:
- Paper 4   Structured questions
- Paper 5   Planning, Analysis and Evaluation

Paper 4 has two sections and you have 2 hours to complete it. Section A consists of structured questions based on the A2 core (Sections I to VI), but may include some material from the AS work. Section B consists of structured questions from Section VII, Gathering and communicating information.

All questions must be answered and you write your answers on lines provided in the question paper.

Paper 5 consists of two questions based on the practical skills of planning, analysis and evaluation. The paper specifically tests practical skills and consequently the work is not necessarily confined to that covered during the A-level course. As for Paper 4, you write your answers on lines provided in the question paper. You have 1 hour 15 minutes to answer this paper.

You can find copies of past papers at:

   http://www.cambridgestudents.org.uk

and click on Subject pages, Physics, AS/AL Physics.

Scientific language

Throughout your physics course, and especially in the examination, it is important to use clear and correct scientific language. Scientists take great care to use language precisely. If researchers do not use exactly the right word when communicating with someone, then what they say could easily be misinterpreted.

Some terms in physics have a clearly defined meaning although they are used in everyday language in a much looser fashion. An example of this is **work**. In physics, you do work when a force moves its point of application in the direction of its line of action. In everyday life, you probably consider yourself to be doing work now, as you read this book!

However, the examiners are testing your knowledge and understanding of physics, not how well you can write in English. They will do their best to understand what you mean, even if some of your spelling and grammar is not correct.

Mathematical skills

You also need to develop your skills in handling equations and mathematical techniques. Physics is a quantitative science and you need a good level of mathematical understanding if you are to fulfil your potential.

Revision

You can download a revision checklist at:

http://www.cambridgestudents.org.uk

and click on Subject pages, Physics, AS/AL Physics. This lists all the learning outcomes, and you can tick them off or make notes about them as your revision progresses.

There are many different ways of revising, and what works well for you may not work for someone else. Look at the suggestions below and try some of them.
- **Revise continually**. Don't think that revision is something you do just before the exam. Life is much easier if you keep revision ticking along right through your physics course. Find 15 minutes a day to look back over work you did a few weeks ago, to keep it fresh in your mind. You will find this helpful when you come to start your intensive revision.
- **Understand it**. Research shows that we learn things much more easily if our brain recognises that they are important and that they make sense. Before you try to learn a topic, make sure that you understand it. If you don't, ask a friend or a teacher, find a different textbook in which to read about it, or look it up on the internet. Work at it until you feel you understand it and then try to learn it.
- **Make your revision active**. Just reading your notes or a textbook will not do any harm but it will not do all that much good. Your brain only puts things into its long-term memory if it thinks they are important, so you need to convince it that they are. You can do this by making your brain *do* something with what you are trying to learn. So, if you are revising the meanings and relationships between electrical quantities or their units, construct a flow diagram to show the relationships. You will learn much more by constructing your own list of bullet points, flow diagram or table than just trying to remember one that someone else has constructed.
- **Fair shares for all**. Don't always start your revision in the same place. If you always start at the beginning of the course, then you will learn a great deal about kinematics and dynamics but not much about nuclear physics. Make sure that each part of the syllabus gets its fair share of your attention and time.

Introduction

- **Plan your time**. You may find it helpful to draw up a revision plan, setting out what you will revise and when. Even if you don't stick to it, it will give you a framework that you can refer to. If you get behind with it, you can rewrite the next parts of the plan to squeeze in the topics you haven't yet covered.
- **Keep your concentration**. It is often said that it is best to revise in short periods, say 20 minutes or half an hour. This is true for many people who find it difficult to concentrate for long periods. But there are others who find it better to settle down for a much longer period of time — even several hours — and really get into their work and stay concentrated without interruptions. Find out which works best for you. It may be different at different times of day. You might be able to concentrate well for only 30 minutes in the morning but are able to get lost in your work for several hours in the evening.
- **Do not assume you know it**. The topics where exam candidates are least likely to do well are often those they have already learned something about at IGCSE or O-level. This is probably because if you think you already know something then you give that a low priority when you are revising. It is important to remember that what you knew for your previous examinations is almost certainly not detailed enough for AS or A2.
- **Must-remember cards.** It is a good idea to write down important equations and definitions on a piece of card, three or four linked equations or definitions on each card. Stick the cards on your bathroom mirror (a few at a time). Read them and learn their contents as you brush your teeth each morning and evening.

The examination

Once you are in the examination room, stop worrying about whether or not you have done enough revision. Concentrate on making the best use of the knowledge, understanding and skills that you have built up through your physics course.

Time

Allow about 1 minute for every mark on the examination paper.

Paper 1 contains 40 multiple-choice questions in 1 hour. Working to the 'one-mark-a-minute' rule, you will have plenty of time to look back over your answers and check any that you weren't quite sure about. Answer every question, even if you only guess at the answer. If you don't know the correct answer, you can probably eliminate one or two of the possible answers, which will increase the chance of your final guess being correct.

In Paper 2, you have to answer 60 marks worth of short-answer questions in 1 hour, which works out at precisely 1 mark per minute. You will probably work a little faster than this, which will give you the chance to check your answers. It is probably worth spending a short time at the start of the examination looking through the whole paper. If you spot a question that you think may take you a little longer than others (for example, a question that has data to analyse), then you can make sure you allow plenty of time for this one.

In Paper 3, you will be working in a laboratory. You have 2 hours to answer questions worth 40 marks. This is much more time per mark than in the other papers, but this is because you will have to do quite a lot of hands-on practical work before you obtain answers to some of the questions. There will be two questions, and your teacher may split the class so that you have to move from one question to the other halfway through the time allowed. It is easy to panic in a practical exam but if you have done plenty of practical work throughout your course, this will help you a lot. Do read through the whole question before you start, and do take time to set up your apparatus correctly and to collect your results carefully and methodically.

Paper 4 contains structured questions. The paper lasts for 2 hours and is worth 100 marks. This gives you the usual 1 mark per minute, plus 20 minutes extra for thinking about difficult questions and checking your work. The paper has two sections. Section A covers sections I to VI of the syllabus. It concentrates on the A2 work but assumes knowledge and understanding of AS material, which might be tested as part of a question. This section is worth 70 marks. Section B covers section VII of the syllabus, 'Applications in physics', and is worth 30 marks. You should split your time so that you spend about 80–85 minutes on Section A and the remainder on Section B.

Paper 5 consists of two questions, each carrying equal marks, which test the practical skills of planning, analysis and evaluation. Although the paper allows 1 hour 15 minutes (75 minutes), the total mark allocation is only 30. This means that you have about two-and-a-half minutes per mark. This is necessary because graph work and data analysis take time. As the paper specifically tests practical skills, the content is not necessarily confined to that covered during the A-level course.

Read the question carefully

This sounds obvious but candidates lose large numbers of marks by not doing it.
- There is often important information at the start of the question that you will need in order to answer the question. Don't just jump straight in and start writing. Start by reading from the beginning of the question. Examiners are usually careful not to give you unnecessary information, so if it is there it is probably needed. You may like to use a highlighter to pick out any particularly important pieces of information at the start of the question.
- Do look carefully at the command words at the start of each question, and make sure that you do what they say. For example if you are asked to *explain* something and you only *describe* it, you will not get many marks — indeed, you may not get any marks at all, even if your description is a good one. You can find the command words and their meanings towards the end of the syllabus.
- Do watch out for any parts of questions that do not have answer lines. For example, you may be asked to label something on a diagram, or to draw a line on a graph, or to write a number in a table. Many candidates miss out these questions and lose significant numbers of marks.

Introduction

Depth and length of answer

The examiners give two useful guidelines about how much you need to write.
- **The number of marks** The more marks, the more information you need to give in your answer. If there are 2 marks, you will need to give at least two pieces of correct and relevant information in your answer in order to get full marks. If there are 5 marks, you will need to write much more. But don't just write for the sake of it — make sure that what you write *answers the question*. And don't just keep writing the same thing several times over in different words.
- **The number of lines** This is not such a useful guideline as the number of marks but it can still help you to know how much to write. If you find your answer won't fit on the lines, then you probably have not focused sharply enough on the question. The best answers are short, precise, use correct scientific language and do not repeat themselves.

Calculations

In all calculations you should show your working clearly. This is good practice and examiners will award marks provided they can follow a calculation, even if you have made an error. If the examiner cannot understand what you have done, or if there is no working, marks cannot be given.

Writing, spelling and grammar

The examiners are testing your knowledge and understanding of physics, not your ability to write English. However, if they cannot understand what you have written, they cannot give you any marks. It is your responsibility to communicate clearly. Do not scribble so fast that the examiner cannot read what you have written. Every year, candidates lose marks because the examiner could not read their writing. It is particularly important to write numbers clearly. If an examiner cannot tell the difference between a 1 and a 7, or a 0 and a 6, the mark cannot be given.

It is also important to include units in your answer, if they are not already printed at the end of the answer cue line. All quantities in physics (except ratios) consist of a number and a unit. You need to know the correct units for each quantity and the correct symbol for those units. A list of the required units is given in the Appendix of the CIE syllabus.

Like spelling, grammar is not taken into consideration when marking your answers — so long as the examiner can understand what you are trying to say. A common difficulty arises when describing the motion of a body in three dimensions. For example, a candidate asked to describe the motion of an electron as it moves through a magnetic field writes, 'The electron accelerates upwards'. Does the candidate mean that the electron moves upwards out of the plane of the paper or upwards towards the top of the paper? If the examiner cannot be sure, you may not be given the benefit of the doubt.

AS
Content
Guidance

I General Physics

General physics
Physical quantities and units

SI units

All quantities in science consist of a number and a unit. There is a system of units used throughout the scientific world known as **SI units**. SI units are based on the units of six base quantities:
- mass — kilogram (kg)
- length — metre (m)
- time — second (s)
- temperature — kelvin (K)
- electric current — ampere (A)
- amount of substance — mole (mol)

Although it is not formally an SI unit, the degree Celsius (°C) is often used as a measure of temperature.

Each of these units has a precise definition. You do not need to remember the details of these definitions.

Derived units

The units of all other quantities are derived from these base units. For example, speed is found by dividing the distance travelled by the time taken. Therefore, the unit of speed is metres (m) divided by seconds (s) which can be written as m/s. At O-level or IGCSE you will probably have written the unit in this way. Now that you are taking your studies a stage further, you should write it as $m\,s^{-1}$.

> **Worked example**
> The unit of force is the newton. What is this in base SI units?
>
> *Answer*
> The newton is defined from the equation:
>   force = mass × acceleration
>   unit of mass = kg
>   unit of acceleration = $m\,s^{-2}$
>
> Insert into the defining equation:
>   units of newton = kg × m × s^{-2}
>   *or* $kg\,m\,s^{-2}$.

Checking homogeneity of equations

If you are not sure if an equation is correct, you can use the units of the different quantities to check it. The units on both sides of the equation must be the same.

Worked example

When a body falls in a vacuum, all its gravitational potential energy is converted into kinetic energy.

By comparing units, show that the equation $mgh = \frac{1}{2}mv^2$ is a possible solution to this equation.

Answer
Write down the units of the quantities on each side of the equation.

Left-hand side; unit of m = kg; unit of g = m s^{-2}; unit of h = m

Right-hand side: unit of $\frac{1}{2}$ = none; unit of m = kg; unit of v = m s^{-1}

Compare the two sides:
  units of mgh = kg × m s^{-2} × m = kg m^2 s^{-2}
  units of $\frac{1}{2}mv^2$ = kg × (m s^{-1})^2 = kg m^2 s^{-2}

Both sides of the equation are identical.

Using standard form

Another way to cope with very large or very small quantities is to use standard form. Here, the numerical part of a quantity is written as a single digit followed by a decimal point, and as many digits after the decimal point as are justified; this is then multiplied by 10 to the required power.

Worked example
The output from a power station is 5.6 GW.

(a)   Express this in watts, using standard form.

(b)   The charge on an electron is 0.000 000 000 000 000 000 16 C.

       Express this in standard form.

Answer
(a)   5.6 GW = 5 600 000 000 W

       5 600 000 000 W = 5.6 × 10^9 W

(b)   0.000 000 000 000 000 000 16 C = 1.6 × 10^{-19} C

       '× 10^{-19}', means that the number, in this case 1.6, is divided by 10^{19}.

An added advantage of using standard form is that it also indicates the degree of precision to which a quantity is measured. This will be looked at in more detail in the section on practical skills.

I General Physics

Multiples and submultiples of base units

Sometimes, the basic unit is either too large or too small. It would not be sensible to use metres to measure the gap between the terminals of a spark plug in a car engine. Instead, you would use millimetres (mm). The prefix 'milli' means divide by 1000 or multiply by 10^{-3}. Similarly, the metre would not be a sensible unit to measure the distance between Cambridge and Hong Kong; you would use kilometres (km). Here, the prefix is 'kilo', which means multiply by 1000 or 10^3.

There are other prefixes which are used to adjust the size of the basic units. Table 1 shows the prefixes that you need to know.

Table 1

Prefix	Symbol	Meaning	
pico-	p	÷ 1 000 000 000 000	× 10^{-12}
nano-	n	÷ 1 000 000 000	× 10^{-9}
micro-	μ	÷ 1 000 000	× 10^{-6}
milli-	m	÷ 1000	× 10^{-3}
centi-	c	÷ 100	× 10^{-2}
deci-	d	÷ 10	× 10^{-1}
kilo-	k	× 1000	× 10^3
mega-	M	× 1 000 000	× 10^6
giga-	G	× 1 000 000 000	× 10^9
tera-	T	× 1 000 000 000 000	× 10^{12}

These are the recognised SI prefixes. The deci- (d) prefix is often used in measuring volume where the decimetre cubed (dm^3) is particularly useful.

Remember that $1\,dm^3$ is 1/1000 of $1\,m^3$ as it is $1/10\,m \times 1/10\,m \times 1/10\,m$.

Making estimates of physical quantities

There are a number of physical quantities, such as the speed of sound in air ($\approx 300\,m\,s^{-1}$), the rough values of which you should be aware. A list of such values is given in appropriate parts of this guide — for example, Table 8, page 68.

Vectors and scalars

A **vector quantity** has magnitude and direction. Examples are force, velocity and acceleration.

A **scalar quantity** has magnitude only. Examples are mass, volume and energy.

When scalars are added the total is simply the arithmetic total. For example, if there are two masses of 2.4 kg and 5.2 kg, the total mass is 7.6 kg.

When vectors are added, their directions must be taken into account. Two forces of 3 N and 5 N acting in the same direction would give a total force of 8 N. However, if they act in opposite directions the total force is (5 − 3) N = 2 N, in the direction of the 5 N force. If they act at any other angle to each other the **triangle of vectors** is used.

International AS and A Level Physics Revision Guide

Constructing a vector diagram

In a vector diagram each vector is represented by a line. The magnitude of the vector is represented by the length of the line and its direction by the direction of the line.

The following rules will help you to draw a triangle of vectors:
1. Choose a suitable scale. Draw a line to represent the first vector (V_1) in both magnitude and direction. Draw a second line, starting from the tip of the first line, to represent the second vector (V_2) in both magnitude and direction.
2. Draw a line from the beginning of the first vector to the end of the second line to complete a triangle.
3. The resultant vector is represented by the length of this line, and its direction.

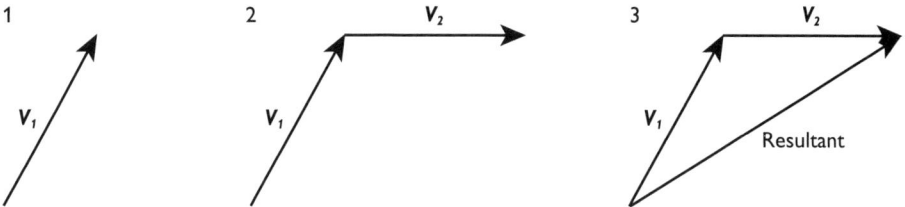

Hints

When you choose a scale, the larger the scale you choose, the greater precision you should achieve in your answer. It is good practice to include your scale on the diagram. When measuring distances use a ruler and when measuring angles use a protractor. Examiners will usually allow a small tolerance.

Worked example

A light aircraft is flying with a velocity relative to the air of 200 km h^{-1} in a direction due north. There is a wind blowing from a direction of 30 degrees north of west at 80 km h^{-1}.

Calculate the velocity of the aircraft relative to the ground.

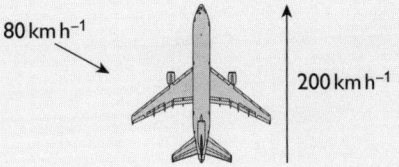

I General Physics

Answer
Draw a vector diagram to a scale 1.0 cm : 40 km h⁻¹.

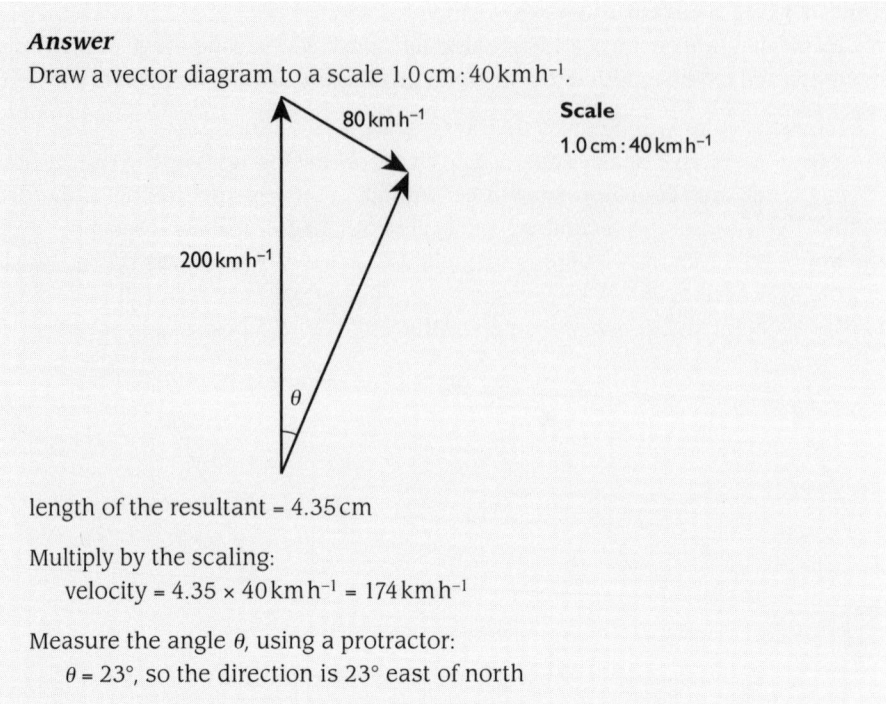

length of the resultant = 4.35 cm

Multiply by the scaling:
velocity = 4.35 × 40 km h⁻¹ = 174 km h⁻¹

Measure the angle θ, using a protractor:
θ = 23°, so the direction is 23° east of north

You will observe that the directions of the original vectors go round the triangle in the same direction (in this example clockwise). The direction of the resultant goes in the opposite direction (anticlockwise). If the original vectors had gone round the triangle in an anticlockwise direction the resultant would have been clockwise.

In simple arithmetic, subtraction is equivalent to the addition of a negative quantity. It is the same when subtracting vectors. The sign of the vector which is to be subtracted is changed. This means that its direction is changed by 180°. This reversed vector is then added to the other vector as described above.

Resolving vectors
Just as it is useful to be able to combine vectors, it is also useful to be able to resolve vectors into components at right angles to each other.

The diagram below shows a vector, **V**, acting at an angle θ to the horizontal:

(a)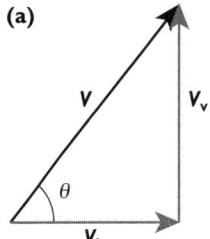

(b)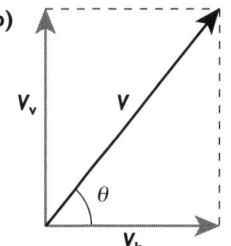

Diagram (a) above shows how the triangle of vectors can be used to show that this vector can be considered to be made up from a vertical (V_v) and a horizontal component (V_h). It is sometimes easier to use a diagram similar to diagram (b) when resolving vectors — this emphasises that the vectors are acting at the same point.

By inspection you can see that $\cos\theta = V_h/V$, therefore

$V_h = V\cos\theta$

Similarly:

$V_v = V\sin\theta$

Worked example
A box of weight 20 N lies at rest on a slope which is at 30° to the horizontal.

Calculate the frictional force on the box up the slope.

Answer
Resolve the weight (20 N) into components parallel to and perpendicular to the slope.

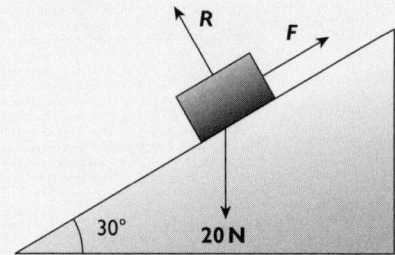 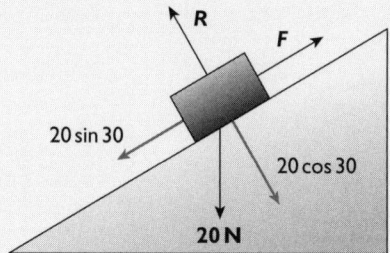

The frictional force, F, is equal to the component of the weight down the slope.

$F = 20\sin 30 = 10\,\text{N}$

Measurement techniques

Physics is a science of measurement and you will need to develop the ability to use a variety of different instruments. Below is a list of instruments and techniques that you should to be able to use. You will have used most of these during the course and this book refers to them where relevant. Nevertheless, it would be good idea to copy the list and, once you feel confident that you can use the instrument proficiently, tick it off.

You should be able to use:
- a ruler, vernier scale and micrometer to measure length
- a lever-arm balance to measure mass

I General Physics

- a spring balance to measure weight
- a protractor to measure angles
- a clock and stopwatch to measure time intervals
- a cathode-ray oscilloscope to measure potential difference
- a cathode-ray oscilloscope with a calibrated time-base to measure time intervals and frequencies
- a thermometer to measure temperature
- an ammeter to measure current
- a voltmeter to measure potential difference
- a galvanometer in null methods

Vernier scales

Rulers can measure to the nearest millimetre, a vernier scale measures to the nearest 1/10 of a millimetre.

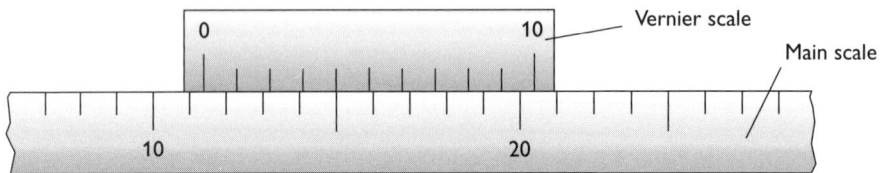

Vernier scale

To read an instrument with a vernier scale:
1 Take the millimetres from the main scale marking before the zero on the vernier scale.
2 Take the next reading from the first vernier mark to coincide with a main scale mark.
3 Add the two readings.

Worked example

What is the reading on the instrument shown in the diagram above?

Answer

main scale reading = 11 mm

vernier reading = 0.4 mm

Calculate the final reading by adding the two readings.

final reading = 11.4 mm

Micrometer scales

Vernier scales can be used to measure to the nearest 1/10 of a millimetre. Micrometers can measure to the nearest 1/100 of a millimetre.

Micrometers have an accurately turned screw thread. When the thimble is turned through one revolution the jaws are opened (or closed) by a predetermined amount. With most micrometers this is 0.50 mm. There will be 50 divisions on the thimble so turning it through one division closes, or opens, the jaws by 0.50 mm divided by 50 = 0.01 mm.

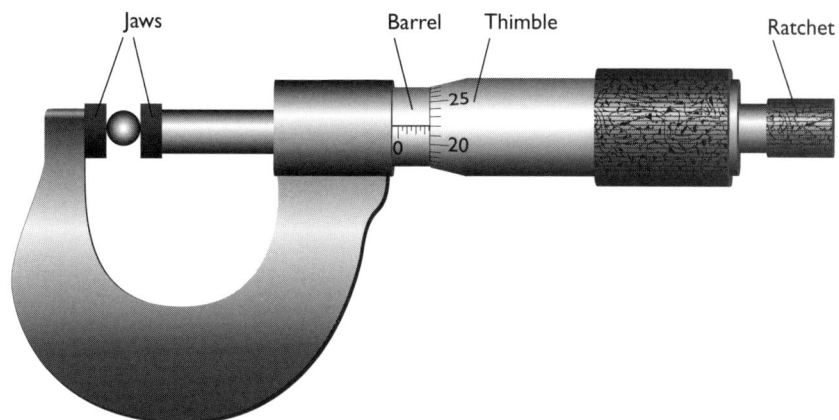

A micrometer screw gauge

To measure the diameter or thickness of an object the jaws are closed, using the ratchet, until they just apply pressure on the object.

To read the micrometer:
1 Take the reading of millimetres and half millimetres from the barrel.
2 Take the reading from the thimble.
3 Add the readings together.

Worked example
What is the reading on the instrument in the diagram above?

Answer
reading on the barrel = 3.5 mm

reading on the thimble = 0.22 mm

Calculate the final reading by adding the two readings.

final reading = 3.72 mm

Cathode-ray oscilloscope

A cathode-ray oscilloscope (c.r.o.) can be used to measure both the amplitude of signals and short time intervals. A potential difference applied to the y-input controls the movement of the trace in a vertical direction, a potential difference applied across the x-input controls the trace in the horizontal direction.

I General Physics

Measurement of potential difference

The y-sensitivity is adjustable and is measured in volts per cm (V cm^{-1}) or volts per division (V div^{-1}).

In the example below, the y-sensitivity is set at 2 V div^{-1}. A d.c. supply is applied across the y-input. No voltage is applied across the x-input. The trace appears as a bright spot.

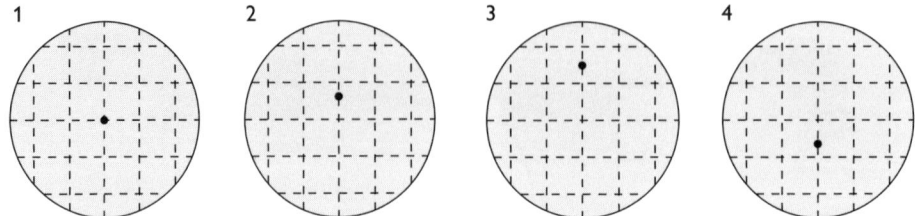

Using a cathode-ray oscilloscope to measure potential difference

In the diagram above:
- Screen 1 shows the cathode-ray oscilloscope with no input.
- Screen 2 shows a deflection of 0.75 of a division. The voltage input is 0.75 × 2 = 1.5 V.
- Screen 3 shows a deflection of 1.5 divisions. The voltage input is 1.5 × 2 = 3.0 V.
- Screen 4 shows a deflection of −0.75 divisions. The voltage input is −0.75 × 2 = −1.5 V, in other words, 1.5 V in the opposite direction.

Measurement of time intervals

To measure time intervals a time-base voltage is applied across the x-input. This drags the spot across the screen, before flying back to the beginning again. The rate at which the time-base voltage drags the spot across the screen can be measured either in seconds per division (s div^{-1}) or divisions per second (div s^{-1}). You *must* check which method has been used.

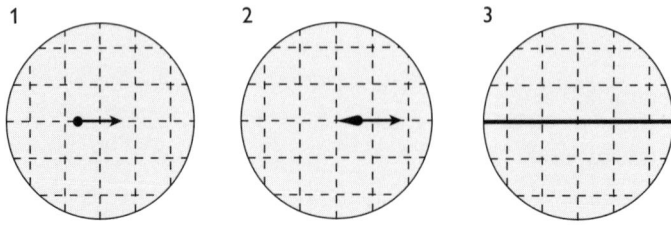

Application of a time-base voltage across the x-input of a cathode-ray oscilloscope

In the diagram above:
- Screen 1 — the spot moves slowly across the screen before flying back to the beginning and repeating the process.
- Screen 2 — with a higher frequency time base, the spot moves across the screen more quickly. The fluorescence on the screen lasts long enough for a short tail to be formed.

- Screen 3 — with a much higher frequency, the fluorescence lasts long enough for the spot to appear as a continuous line.

If successive pulses are applied to the y-plate while the time-base voltage is applied, the trace might appear as in the diagram below:

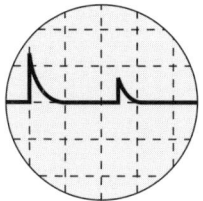

Using a cathode-ray oscilloscope to measure time intervals

The time interval between the pulses can be calculated by multiplying the number of divisions between the two pulses by the time base.

Worked example
A survey ship sends a pulse of sound down to the seabed and the echo is detected. The two pulses are shown in the diagram above with the cathode-ray oscilloscope time base being set at $50\,\text{ms div}^{-1}$. Calculate the depth of the sea, given that the speed of sound in water is $1500\,\text{ms}^{-1}$.

Answer
time interval between pulses = number of divisions × time base = 2.5 × 50
= 125 ms = 0.125 s.

speed = distance/time

Therefore:
 distance = speed × time = 1500 × 0.125 = 187.5 m
 depth of the water = 187.5/2 ≈ 94 m

Calibration curves

You may well use sensors whose output is not proportional to the quantity you are attempting to measure. A good example is the output from a thermocouple thermometer, which you will have met in your pre-AS course. The example shows how you can use a calibration curve to use this type of instrument.

Worked example
The graph on page 24 shows the calibration curve for a thermocouple used to measure temperatures from 0 to 250°C.

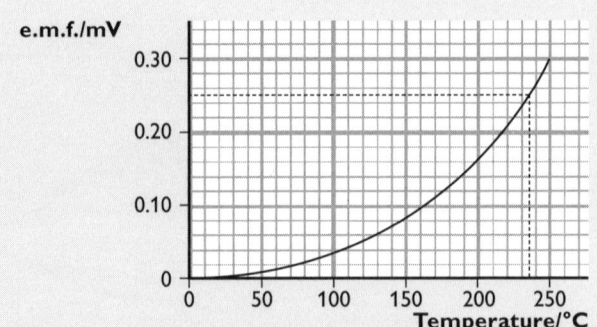

Deduce the temperature when the thermocouple produces an e.m.f. of 0.250 mV.

Answer
Draw a horizontal line from 0.250 mV on the *y*-axis to the curve.

Draw a vertical line from where this line intersects with the curve to cut the *x*-axis.

This intersects at 240°C, hence the required temperature is 240°C.

Newtonian mechanics

Kinematics

Definitions of quantities

Displacement is the distance an object is from a fixed reference point in a specified direction.

Displacement is a vector quantity. It has both magnitude and direction.

Speed is the distance travelled per unit time.
Speed is a scalar quantity. It refers to the total distance travelled.

Velocity is the change in displacement per unit time.
Velocity is a vector quantity, being derived from displacement — not the total distance travelled.

Acceleration is the rate of change of velocity.
Acceleration is a vector quantity. An acceleration in the direction in which a body is travelling will increase its velocity. An acceleration in the opposite direction to which a body is travelling will decrease its velocity. An acceleration at angle 90° to which a body is travelling will change the direction of the velocity but will not change the magnitude of the velocity.

Equations linking the quantities

$v = \frac{\Delta s}{\Delta t}$, where v is the velocity, Δs is the change of displacement in a time Δt.

In general, the symbol Δ means 'change of', so Δs is the change in displacement and Δt is the change in time.

$a = \frac{\Delta v}{\Delta t}$, where a is the acceleration, Δv is the change in velocity in time Δt.

Units

Speed and velocity are measured in metres per second ($m\,s^{-1}$)

Acceleration is the change in velocity per unit time; velocity is measured in metres per second ($m\,s^{-1}$) and time is measured in seconds (s), which means that the acceleration is measured in metres per second every second ($m\,s^{-1}$ per s) which is written as $m\,s^{-2}$.

Worked example
A toy train travels round one circuit of a circular track of circumference 2.4 m in 4.8 s.

Calculate:
(a) the average speed
(b) the average velocity

Answer
(a) average speed $= \frac{\Delta x}{\Delta t} = \frac{2.4\,(m)}{4.8\,(s)} = 0.50\,m\,s^{-1}$

(b) s is the displacement, which after one lap is zero. The train finishes at the same point that it started.

Hence:

average $v = \frac{\Delta s}{\Delta t} = \frac{0\,(m)}{4.8\,(s)}$ and $v = 0\,m\,s^{-1}$

It is good practice to include units in your calculations, as shown in this example — it can help to avoid mistakes with multiples of units. It can also help you to see if an equation does not balance. In this book, however, units are only included in the final quantity to make the equation clear.

The example above demonstrates the difference between speed and velocity.

Worked example
A car travels 840 m along a straight level track at constant speed of $35\,m\,s^{-1}$. The driver then applies the brakes and the car decelerates to rest at a constant rate in a further 7.0 s.

II Newtonian mechanics

Calculate:
(a) the time for which the car is travelling at constant speed
(b) the acceleration of the car when the brakes are applied

Answer
(a) $v = \frac{\Delta s}{\Delta t}$, $35 = \frac{840}{\Delta t}$, $\Delta t = \frac{840}{35}$

$t = 24\,\text{s}$

(b) $a = \frac{\Delta v}{\Delta t} = \frac{0 - 35}{7.0}$

$a = -5.0\,\text{m s}^{-2}$

The minus sign shows that the velocity decreases rather than increases. It is also worth noting that the given quantities in the question are to two significant figures. Therefore, the answer should also be recorded to two significant figures.

Graphs

Graphs give a visual picture of the manner in which one variable changes with another. Looking at motion graphs can help us to see what is happening over a period of time.

Displacement–time graphs

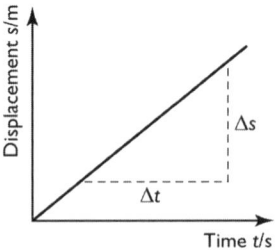

This graph shows the displacement of a body which increases uniformly with time. This shows constant velocity. The magnitude of the velocity is equal to the gradient of the graph.

$v = \text{gradient} = \frac{\Delta s}{\Delta t}$

Hint

When you measure the gradient of a graph, use as much of the graph as possible. This will reduce the percentage error in your calculation.

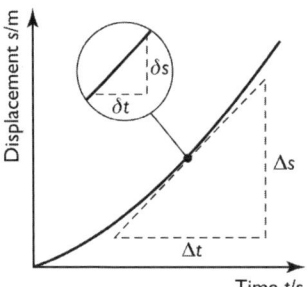

This graph shows an example of body's velocity steadily increasing with time. To find the velocity at a particular instant (the instantaneous velocity) we draw a tangent to the graph at the relevant point and calculate the gradient of that tangent.

Velocity–time graphs

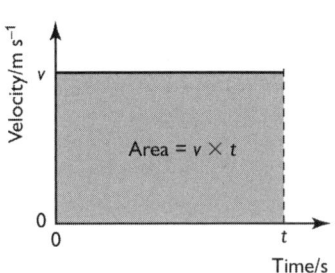 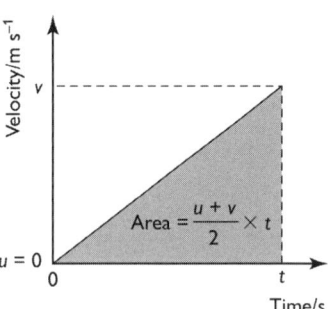

The left-hand graph shows a body moving with a constant velocity; the right-hand graph shows that the velocity of the body is increasing at a constant rate — it has constant **acceleration.**

The gradient of a velocity–time graph is the change in velocity divided by the time taken. It is equal to the magnitude of the acceleration.

$$a = \text{gradient} = \frac{v - u}{t_2 - t_1} = \frac{\Delta v}{\Delta t}$$

Displacement from a velocity–time graph
The displacement is equal to the area under a velocity–time graph; this can be clearly seen in the left-hand graph above. The shaded area is a rectangle and its area is equal to:

length × height = velocity × time

The right-hand graph shows changing velocity and the distance travelled is the average velocity multiplied by the time. For constant acceleration from zero velocity this is half the maximum velocity multiplied by the time — the area of a triangle.

Worked example

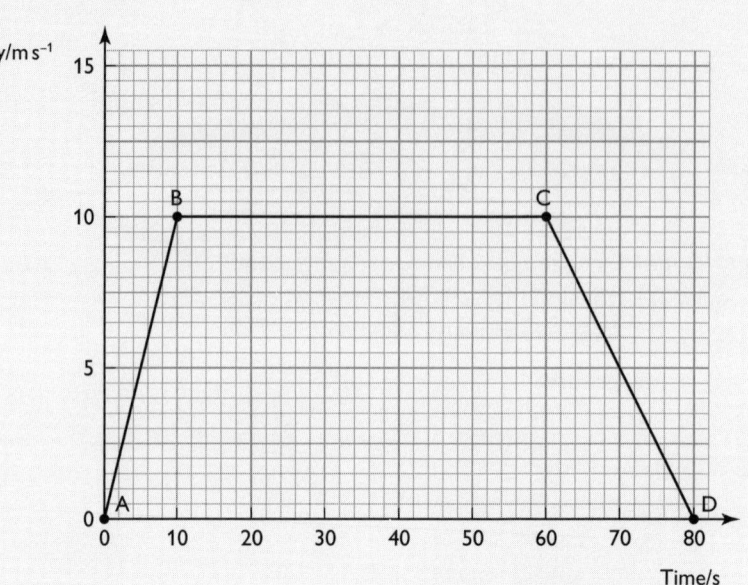

The graph shows the motion of a cyclist as she travels from one stage to the next in a race.

Calculate
- **(a)** the acceleration from A to B
- **(b)** the maximum speed of the cyclist
- **(c)** the total distance the cyclist travels
- **(d)** the acceleration from C to D

Answer

(a) acceleration = gradient = $\frac{10-0}{10-0}$

= $1.0 \, m\,s^{-2}$

(b) The maximum speed can be read directly from the graph. It is $10 \, m\,s^{-1}$.

(c) distance travelled = area under the graph

= (½ × 10 × 10) + (10 × 50) + (½ × 10 × 20)

= 650 m

Deriving equations of uniformly accelerated motion

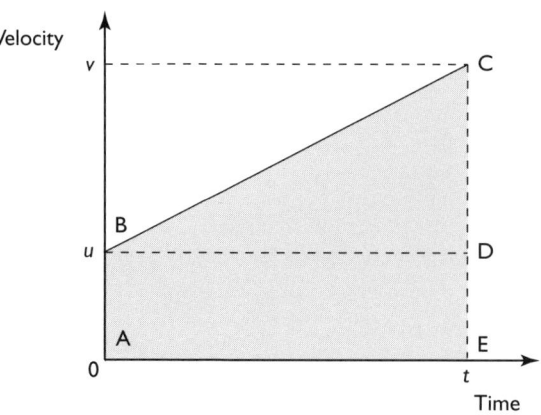

The graph shows the motion of a body which has accelerated at a uniform rate, from an initial velocity u to a final velocity v in time t.

Equation 1
The acceleration of the body:

$$a = \frac{v - u}{t}$$

Rearranging this equation gives:

$$v = u + at$$

Equation 2
The distance travelled s by the body can be found in two ways:

s = average velocity × time

$$s = \frac{v + u}{t} t$$

Equation 3
distance travelled = area under the graph

s = area of rectangle ABDE + area of the triangle BCD

$s = ut + ½(v - u)t$

$(v - u)/t = a$

Therefore, $s = ut + ½at^2$

Equation 4
A fourth equation is needed to solve problems in which the time and one other variable are not known.

Equation 1 rearranges to $t = (v - u)/a$

Substitute this in Equation 2:

II Newtonian mechanics

$$S = \frac{v+u}{2} \times \frac{v-u}{a}$$

$$S = \frac{v^2 - u^2}{2a}$$

Rearranging gives:

$$v^2 = u^2 + 2as$$

Summary

The equations of uniformly accelerated motion are:
- $v = u + at$
- $s = \frac{v+u}{2}t$
- $s = ut + \frac{1}{2}at^2$
- $v^2 = u^2 + 2as$

These equations of motion can only be used if there is constant acceleration (including constant deceleration and zero acceleration) for the whole part of the journey that is being considered.

Using the equations of uniformly accelerated motion

A common type of problem you might be asked to analyse is the journey of a vehicle between two fixed points.

> **Worked example**
> During the testing of a car, it is timed over a measured kilometre. In one test it enters the timing zone at a velocity of $50\,\text{m s}^{-1}$ and decelerates at a constant rate of $0.80\,\text{m s}^{-2}$.
>
> Calculate:
> (a) the velocity of the car as it leaves the measured kilometre
>
> (b) the time it takes to cover the measured kilometre
>
> **Answer**
> (a) $u = 50\,\text{m s}^{-1}$
> $s = 1.0\,\text{km} = 1000\,\text{m}$
> $a = -0.80\,\text{m s}^{-2}$
> $v = ?$
>
> Required equation: $v^2 = u^2 + 2as$
> Substitute the relevant values and solve the equation:
> $v^2 = 50^2 + 2 \times (-0.80) \times 1000 = 2500 - 1600 = 900$
> $v = 30\,\text{m s}^{-1}$

(b) Required equation: $v = u + at$

Substitute in the relevant variables:
$30 = 50 - (0.80 \times t)$
$t = (50 - 30)/0.8$
$= 25\,\text{s}$

Hint

It might seem tedious writing out all the quantities you know and the equation you are going to use. However, this will mean that you are less likely to make a careless error and, if you do make an arithmetic error, it helps the examiner to see where you have gone wrong, so that some marks can be awarded.

Two common mistakes in this type of question are:
- forgetting deceleration is a negative acceleration
- forgetting to convert kilometres to metres

Analysing the motion of a body in a uniform gravitational field

The equations of uniformly accelerated motion can be used to analyse the motion of a body moving vertically under the influence of gravity. In this type of example it is important to call one direction positive and the other negative and to be consistent throughout your calculation. The following example demonstrates this.

Worked example

A boy throws a stone vertically up into the air with a velocity of $6.0\,\text{m}\,\text{s}^{-1}$. The stone reaches a maximum height and falls into the sea, which is 12 m below the point of release.

Calculate the velocity at which the stone hits the water surface.

(acceleration due to gravity = $9.8\,\text{m}\,\text{s}^{-2}$)

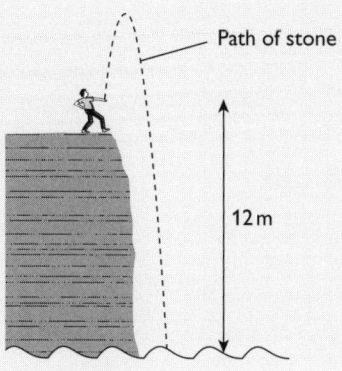

II Newtonian mechanics

Answer
$u = 6.0\,\text{m s}^{-1}$
$a = -9.8\,\text{m s}^{-2}$
$s = -12\,\text{m}$
$v = ?$

Required equation: $v^2 = u^2 + 2as$

$v^2 = 6.0^2 + [2 \times (-9.8) \times (-12)] = 36 + 235 = 271$

$v = \pm 16.5\,\text{m s}^{-2}$

In this example, upwards has been chosen as the positive direction; hence u is $+6.0\,\text{m s}^{-1}$. Consequently, the distance of the sea below the point of release (12 m) and the acceleration due to gravity (10 m s^{-2}) are considered negative because they are both in the downwards direction.

The final velocity of the stone is also in the downwards direction. Therefore, it should be recorded as $-16.5\,\text{m s}^{-2}$.

It is also worth noting that air resistance on a stone moving at these speeds is negligible and can be ignored.

Mass and weight

The mass of a body is the amount of matter in the body; its weight is the gravitational force on the body. Thus mass is independent of the position of a body, whereas weight varies according to the gravitational field in which the body lies. In general, the two are connected by the equation:

$W = mg$

where W = weight, m = mass and g = gravitational field strength (or acceleration of free fall).

The gravitational field strength near the surface of the Earth is 9.8 N kg^{-1}. Therefore, a mass of about 100 g (0.1 kg) has a weight of just less than 1 N (0.98 N) on the Earth's surface. Its weight on the Moon is only 0.16 N because the gravitational field strength on the Moon is only about 1/6 of that on Earth.

Acceleration of free fall

In the absence of air resistance, all bodies near the Earth fall with the same acceleration. This is known as the acceleration of free fall. Similarly, bodies near any other planet will fall with equal accelerations. However, these accelerations will be different from those near the Earth. This is explored further in the section on dynamics.

Measurement of the acceleration of free fall

The diagrams below show apparatus that can be used to measure the acceleration of free fall.

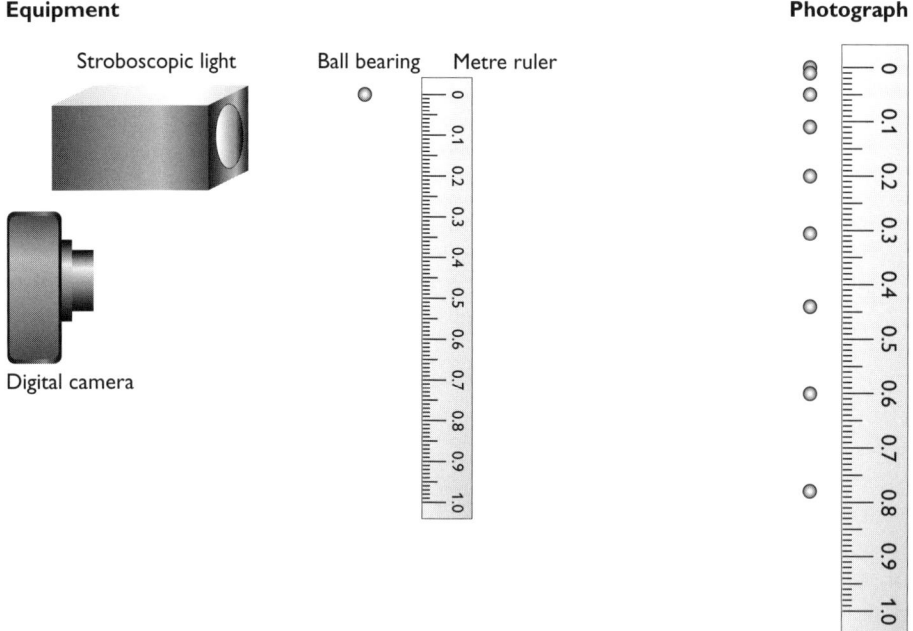

The stroboscopic light flashes at a fixed frequency and the shutter of the camera is held open. This results in a photograph that shows the position of the ball in successive time intervals as in the diagram. In this example the stroboscopic light was set to flash at 20 Hz. In Table 2, the third column shows the distance travelled by the ball in each time interval and the fourth column shows the average speed during each interval.

Table 2

Time/s	Position/m	Distance travelled/m	Speed/m s^{-1}
0.00	0.00	0.00	0.0
0.05	0.01	0.01	0.2
0.10	0.05	0.04	0.8
0.15	0.11	0.06	1.2
0.20	0.20	0.09	1.8
0.25	0.30	0.11	2.2
0.30	0.44	0.14	2.8
0.35	0.60	0.16	3.2
0.40	0.78	0.18	3.6

II Newtonian mechanics

The graph of speed against time is plotted. Acceleration is equal to the gradient of this graph:

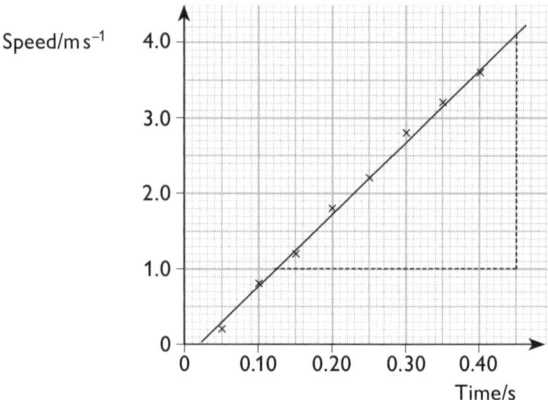

$$a = \frac{4.1 - 1.0}{0.45 - 0.13} = 9.7 \, \text{m s}^{-2}$$

Effect of air resistance

When you kick a football or hit a tennis ball you will be aware of the effect of air resistance. Air resistance affects all moving bodies near the Earth's surface, including the motion of falling bodies. Air resistance depends on the shape of a body and also on the speed at which the body travels. The resistance on a streamlined body is less than on a less streamlined body. Car manufacturers spend a lot of time and money researching the best shape for a car so as to reduce air resistance.

Air resistance, or drag, increases as the velocity of a body increases. As a falling body accelerates, the drag force increases. Therefore, the resultant force on it will decrease, meaning that the acceleration decreases. When the drag force is equal to the gravitational pull on the body it will no longer accelerate, but fall with a constant velocity. This velocity is called the **terminal velocity**.

The graph below shows how the velocities of a shuttlecock and of a tennis ball change as they fall from rest.

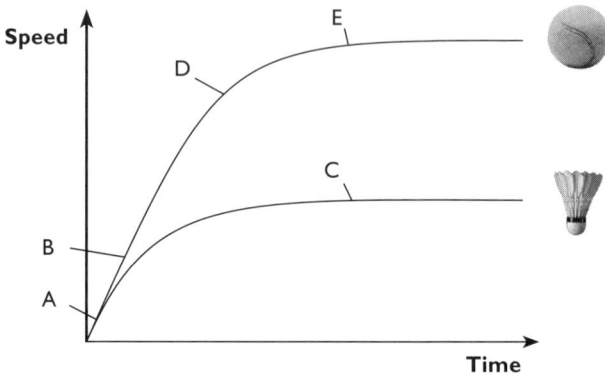

In the graph above:
- at **point A** the air resistance (or drag force) is negligible and both the shuttlecock and the tennis ball fall with the same acceleration, g
- at **point B** the air resistance (compared with the weight of the ball) remains small and it continues to fall with the same acceleration. The shuttlecock has a much smaller weight than the ball and the air resistance on it is significant compared to its weight, so its acceleration is reduced
- at **point C** the air resistance is equal to the weight of the shuttlecock. It no longer accelerates and falls with its terminal velocity
- at **point D** the air resistance on the ball is now significant and its acceleration is reduced
- at **point E** the air resistance is equal to the weight of the ball and it falls with its terminal velocity

Motion in two dimensions

Consider a small object in outer space travelling at a velocity v. A force acts on it at right angles to its velocity for a short time.

The magnitude of the velocity will not change but there will be a change of direction due to the force. The change in direction will be in the direction of the force. This principle is used to make course direction changes when spacecraft are sent to different planets.

If the force is large enough and is maintained for some time the particle will continue to change its direction and will travel in a circular path as shown in the diagram below.

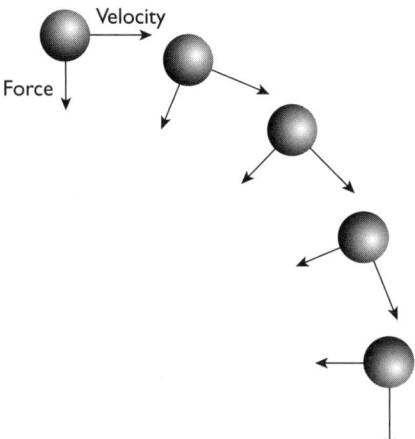

Another example of a body travelling under a force that is not parallel to the body's motion is when the body is travelling in a uniform field. A simple example of this is a cricket ball thrown horizontally near the Earth's surface. If the effect of air resistance is ignored the ball will describe a parabolic path. The point to recognise here is that the horizontal and vertical components of the velocity can be considered separately. The horizontal component remains constant as there is no force (and hence no acceleration) in that direction. The vertical component will gradually increase in exactly the same way as the velocity of a ball in free fall would increase.

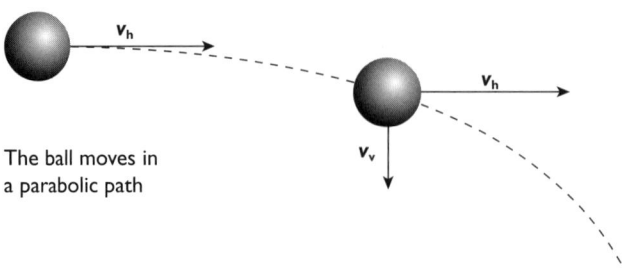

The ball moves in a parabolic path

Worked example

A golf ball is hit so that it leaves the club face at a velocity of $45\,\mathrm{m\,s^{-1}}$ at an angle of $40°$ to the horizontal.

Calculate

(a) the horizontal component of the velocity

(b) the vertical component of the velocity

(c) the time taken for the ball to reach its maximum height

(d) the horizontal distance travelled when the ball is at its maximum height

(Ignore the effects of air resistance and spin on the ball.)

Answer

(a) $v_h = v\cos\theta = 45\cos 40 = 34.5\,\mathrm{m\,s^{-1}}$

(b) $v_v = v\sin\theta = 45\sin 40 = 28.9\,\mathrm{m\,s^{-1}}$

(c) In the vertical direction the motion can be considered to be that of a ball thrown vertically upwards decelerating under the effect of gravity. At the top of the flight the vertical velocity will be zero. Use the equation:

$v = u + at$
$0 = 28.9 + (-9.8)t$
$t = 2.95 \approx 3.0\,\mathrm{m\,s^{-1}}$

(d) The horizontal component of the velocity remains constant throughout the flight.

$s = ut = 34.5 \times 2.95 = 102 \approx 100\,\mathrm{m}$

If the effects of air resistance and spin are ignored the flight path would be symmetrical. This means that if the ball were hit on a level field, it would travel a total horizontal distance of 200 m before bouncing.

Dynamics

Newton's laws of motion

Newton's laws are the basis on which classical mechanics was built. Many of the ideas are seen as self-evident today but were revolutionary when Newton first developed his ideas.

Before discussing Newton's laws of motion in detail you need to understand the concept of momentum.

Momentum

Linear momentum (p) is defined as the product of mass and velocity:

$$p = mv$$

The unit of momentum is $kg\,m\,s^{-1}$. It is formed by multiplying a vector by a scalar and is, therefore, a vector itself. This means, for example, that a body of mass 2 kg travelling at $3\,m\,s^{-1}$ has a momentum of $6\,kg\,m\,s^{-1}$. A body of the same mass travelling at the same speed but in the opposite direction has a momentum of $-6\,kg\,m\,s^{-1}$. It is important when you consider interactions between bodies that you understand the vector nature of momentum.

> **Worked example**
> Calculate the momentum of a cruise liner of mass 20 000 tonnes when it is travelling at $6.0\,m\,s^{-1}$ (1 tonne = 1000 kg).
>
> *Answer*
> Convert the mass to kg: 20 000 t = 20 000 × 1000 kg = 20 000 000 kg
>
> $p = mv = 20\,000\,000 \times 6.0 = 120\,000\,000\,kg\,m\,s^{-1} = 1.2 \times 10^8\,kg\,m\,s^{-1}$

First law

A body will remain at rest or move with constant velocity unless acted on by a resultant force.

The first part of this law is relatively straightforward; we do not expect an object to move suddenly for no reason. The second part requires a little more thought. A golf ball putted along level ground will gradually slow down, as will a cyclist freewheeling along a level path. In both these cases frictional forces act in the opposite direction to the velocity of the body and cause it to decelerate. When we observe motion on the Earth we cannot eliminate friction and we 'learn' (falsely) that a force is needed to keep bodies moving. In practice, we only need that force to overcome frictional forces. If you think of a rock moving through outer space, there is no force on it — yet it will continue moving in a straight line forever, or until it encounters another body, perhaps in another galaxy.

II Newtonian mechanics

Second law

A resultant force acting on a body will cause a change in momentum in the direction of the force. The rate of change of momentum is proportional to the magnitude of the force.

The first law describes what happens when there is no force on a body, this second law explains what happens when there is a force on a body. Not only that, it defines what a force is: something that tends to cause a change in momentum of a body. From this law we can write:

$$F \propto \frac{\Delta p}{\Delta t}$$

The constant of proportionality defines the size of the unit of force. The newton is defined by making the constant equal to 1, when momentum is measured in $kg\,m\,s^{-1}$ and time is measured in s.

Hence:

$$F = \frac{\Delta p}{\Delta t}$$

You see from this equation that force is measured in $kg\,m\,s^{-2}$. $1\,kg\,m\,s^{-2}$ is called 1 N (newton).

Worked example
A golf ball of mass 45 g is putted along a level green with an initial velocity of $4.0\,m\,s^{-1}$. It decelerates at a constant rate and comes to rest after 3.0 s.

Calculate the frictional force on the ball.

Answer
Convert the mass to kg: $45\,g = 45/1000\,kg = 0.045\,kg$

initial momentum $= 0.045 \times 4 = 0.18\,kg\,m\,s^{-1}$

final momentum $= 0$

change in momentum $= -0.18\,kg\,m\,s^{-1}$

$$F = \frac{\Delta p}{\Delta t} = \frac{-0.18}{3.0} = -0.060\,N$$

The minus sign in the answer shows that the force is acting in the opposite direction to the initial velocity.

Acceleration of a constant mass
In many situations, including the previous worked example, the mass of the body on which the force is applied remains constant (or nearly constant). Consider the basic equation:

$$F = \frac{\Delta p}{\Delta t}$$

Now $\Delta p = \Delta(mv)$ and if m is constant this can be rewritten as $p = m\Delta v$.

Therefore: $F = \dfrac{m\Delta v}{\Delta t}$

but $\dfrac{\Delta v}{\Delta t}$ = acceleration

so, **F = ma**

The previous example could be solved using this equation, rather than using the rate of change of momentum.

> **Worked example**
> A car of mass 1.2 tonnes accelerates from $5\,\text{m s}^{-1}$ to $30\,\text{m s}^{-1}$ in $7.5\,\text{s}$.
>
> Calculate the average accelerating force on the car.
>
> **Answer**
> acceleration = change in velocity/time taken = $(30 - 5)/7.5 = 3.3\,\text{m s}^{-2}$
>
> convert the mass to kilograms: $1.2\,\text{t} = 1200\,\text{kg}$
>
> force = mass × acceleration = $1200 × 3.3 = 4000\,\text{N}$

This equation also gives a deeper insight into the concept of mass. You can see that the greater the mass of a body, the harder it is to change its uniform velocity. You begin to see that mass is a measure of this 'reluctance to change', or **inertia**.

Third law

Newton's third law looks at the interaction between two bodies.

If body A exerts a force on body B then body B will exert a force on body A of equal magnitude but in the opposite direction.

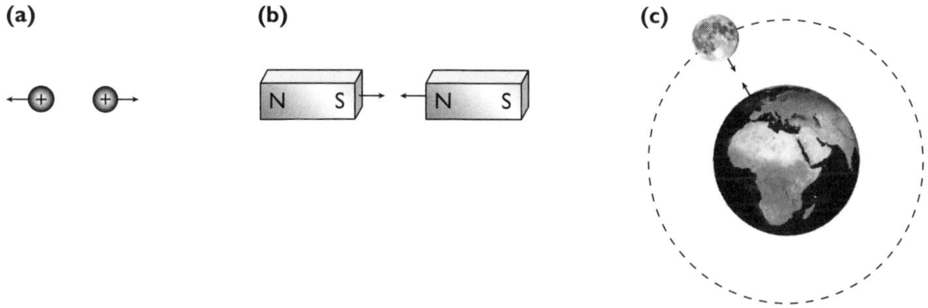

(a)   (b)   (c)

(a) Two protons repel each other
(b) Two magnets attract each other
(c) The Earth and Moon attract each other

The examples in the diagram above show forces on two bodies of roughly equal size; it is easy to appreciate that the forces in each example are of equal size. However, it is also true with objects of very different sizes. For example, when you jump off a wall there is a gravitational pull on you from the Earth that pulls you down towards the ground. What you do not think about is that you also pull the Earth upwards towards you with an equal sized force! Of course, the movement of the Earth is negligible because it is so much more massive than you are — but the force is still there.

The child is pulled down by the Earth with a force, W

The Earth is pulled up by the child with a force, W

Conservation of momentum

One of the useful results that can be developed from Newton's third law is that momentum is conserved in any interaction. This means that the total momentum of a closed system (that is a system on which no external forces act) is the same after an interaction as before.

Consider two bodies that move towards each other, as in the diagram below, and then stick to each other after the collision.

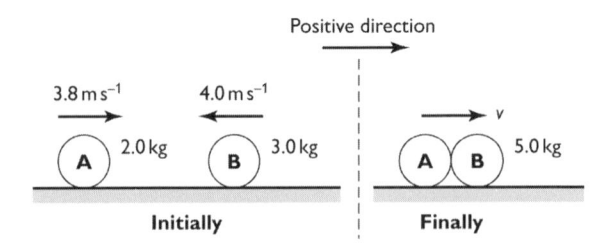

total momentum before the collision = total momentum after the collision

If we consider the positive direction to be from left to right:

$(2.0 \times 3.8) + (3.0 \times -4.0) = 5v$

$-4.4 = 5v$

$v = -0.88 \, m\,s^{-1}$

The negative sign means that the velocity after the collision is from right to left.

A formal statement of the law is as follows:

The total momentum of a closed system before an interaction is equal to the total momentum of that system after the interaction.

Three types of interaction

Elastic: in an elastic interaction not only is the momentum conserved, the kinetic energy is also conserved.

On the macroscopic scale this is rare. However many interactions do approximate to being perfectly elastic and the mathematics of an elastic interaction can be used to model these. On the microscopic scale, for example, the collision between two charged particles such as protons can be considered to be elastic.

It is worth noting that in any perfectly elastic collision the relative speed of approach before the interaction is equal to the relative speed of separation after the interaction. A good example of this is the nearly elastic interaction of a golf ball being struck by the much more massive club. The ball will leave the club at twice the speed at which the club approaches the ball.

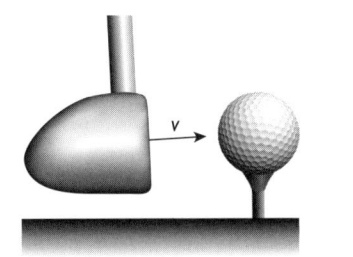

The golf club approaces the ball at a velocity of v.

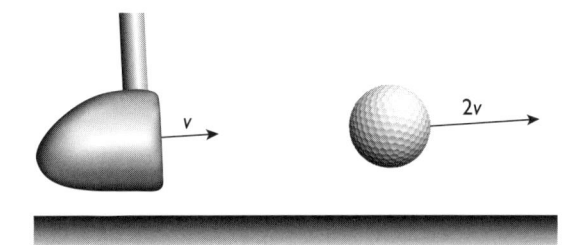

The club continues to move at a velocity of (very nearly) v. The ball moves off at a speed of (nearly) $2v$. The speed of separation of the ball from the club is equal to the speed of approach of the club to the ball.

Inelastic: in an inelastic collision some of the initial kinetic energy is converted into other forms, such as sound and internal energy. The kinetic energy is less after the collision than before it. As in all collisions, momentum is conserved. There are numerous examples and degrees of inelastic collision — from nearly perfectly elastic, such as one billiard ball striking another, to two bodies sticking together, such as two identical trolleys colliding and sticking together as shown in the diagram below:

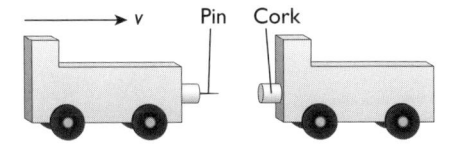

One trolley moves towards a second identical stationary trolley with a speed v.

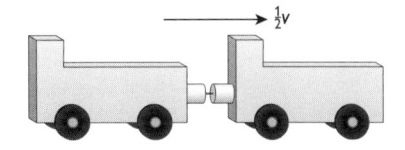

The two trolleys stick together and move off with a combined speed of $\frac{1}{2}v$.

Explosive: in an explosive interaction potential energy is converted to kinetic energy; the kinetic energy is greater after the interaction than before. An example of this is the emission of an alpha particle from a radioactive nucleus. There is no change in the total momentum of the system in this type of interaction. Initially, there is zero momentum. After the interaction, the momentum of the alpha particle in one direction is of the same magnitude as the momentum of the recoiling daughter nucleus in the other direction. Therefore, the total momentum remains zero.

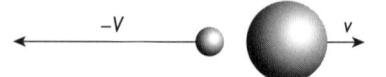

Stationary radium nucleus

Alpha particle emitted at high speed. The (much more massive) daughter nucleus recoils, at a much lower speed.

Worked example

A glider of mass 0.20 kg on an air track is moving at 3.6 m s^{-1} towards a second glider of mass 0.25 kg, which is moving at 2.0 m s^{-1} in the opposite direction. When the two gliders collide they stick together.

(a) Calculate their joint velocity after the collision.

(b) Show that the collision is inelastic.

Answer

(a) momentum before the collision = (0.20 × 3.6) + (0.25 × −2.0)
= 0.22 kg m s^{-1}

momentum after the collision = (0.20 + 0.25)v = 0.45v, where v is the velocity of the two gliders after the collision

momentum after the collision = momentum before the collision
0.22 = 0.45v
v = 0.49 m s^{-1}

(b) kinetic energy before the collision = (½ × 0.2 × 3.6^2) + (½ × 0.25 × 2.0^2)
= 1.3 + 0.5
= 1.8 J

Kinetic energy after the collision = (½ × 0.45 × 0.49^2) = 0.054 J

The kinetic energy after the collision is less than the kinetic energy before the collision, therefore the collsion is inelastic.

Forces

Types of force

In your work before AS you will have met the idea of a force being a push or a pull. You should now recognise the slightly more sophisticated idea that a force causes, or tends to cause, a change in the velocity of a body.

You have met various types of force already. Here is a list of the types of force with which you should be familiar:
- gravitational force
- electric force
- upthrust or buoyancy forces
- frictional and viscous forces

Gravitational forces

A mass in a gravitational field experiences a force. We have already seen that the size of the force depends on the strength of the gravitational field and the mass of the object:

$$F = mg$$

where F = force (or weight), m = mass of the body and g = gravitational field strength.

Near the Earth's surface (or any planetary-sized body) the gravitational field is uniform. Therefore, the gravitational force is the same wherever the body is placed near the planet's surface. Consequently, the body will fall with a constant acceleration (ignoring air resistance). Near the Earth's surface the gravitational field is approximately $9.8\,\text{N}\,\text{kg}^{-1}$. This will cause any object to fall with an acceleration of $9.8\,\text{m}\,\text{s}^{-2}$. The gravitational field near the Moon is $1.6\,\text{N}\,\text{kg}^{-1}$. Consequently, an object near the Moon's surface will fall towards the Moon's surface with an acceleration of $1.6\,\text{m}\,\text{s}^{-2}$.

Electric forces

A charged object will experience forces due to other charged objects nearby. The behaviour of a charged object in a uniform electric field is investigated on pages 80–82.

Upthrust or buoyancy forces

Bodies wholly or partly immersed in fluids experience an upthrust due to the slightly different pressures exerted on their lower and upper surfaces. This is explored further on pages 57–58.

Frictional and viscous forces

Frictional forces have already been discussed. The term 'friction' is usually applied where there is resistance to motion due to contact between two solids. It arises because no two surfaces are perfectly smooth and the lumps in them tend to interlock when there is relative movement between the bodies:

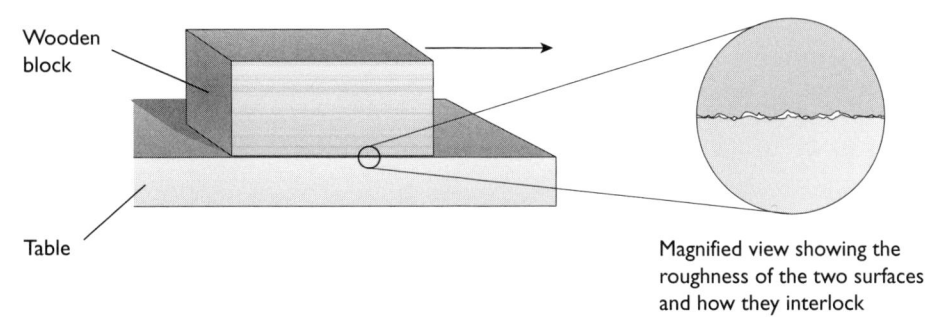

Wooden block

Table

Magnified view showing the roughness of the two surfaces and how they interlock

II Newtonian mechanics

The term 'viscous' tends to be used when fluids (liquids and gases) are involved. It is the difference in viscosity that makes water flow much more quickly than oil. Similarly, the viscous forces on a body travelling through oil are much larger than those on an object travelling through water. Gases tend to produce far less viscous drag than liquids. Even so, at high speeds the viscous drag on cars and aircraft is significant.

Forces in equilibrium

In the section on vectors you met the idea of using vector diagrams to add vectors acting at different angles, and you also met the concept of resolving vectors into their component parts. Vector diagrams can also be used when a body is in equilibrium.

Consider a lamp of weight W pulled to one side with a horizontal force F so that it makes an angle θ with the vertical:

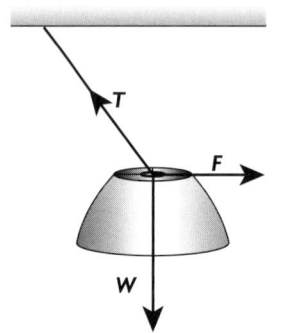

The diagram shows the forces acting on the lamp. T is the tension in the flex.

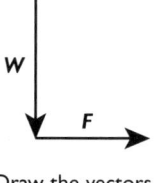

Draw the vectors W and F.

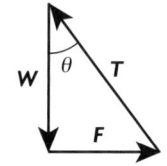

Complete the triangle. The third side represents the tension T.

Note the difference between this and using the parallelogram of forces to find the resultant of two forces. This is a closed triangle, with all the arrows going the same way round the triangle. This shows that the sum of these three forces is zero and that the body is in equilibrium.

Worked example
A helium balloon is tethered to the ground using a cable that can withstand a maximum force of 10.0 kN before breaking. The net upwards force on the balloon due to its buoyancy is 8.0 kN.

Calculate the maximum horizontal force the wind can produce on the balloon before the cable snaps, and the angle the cable makes with the vertical when this force is applied.

Answer

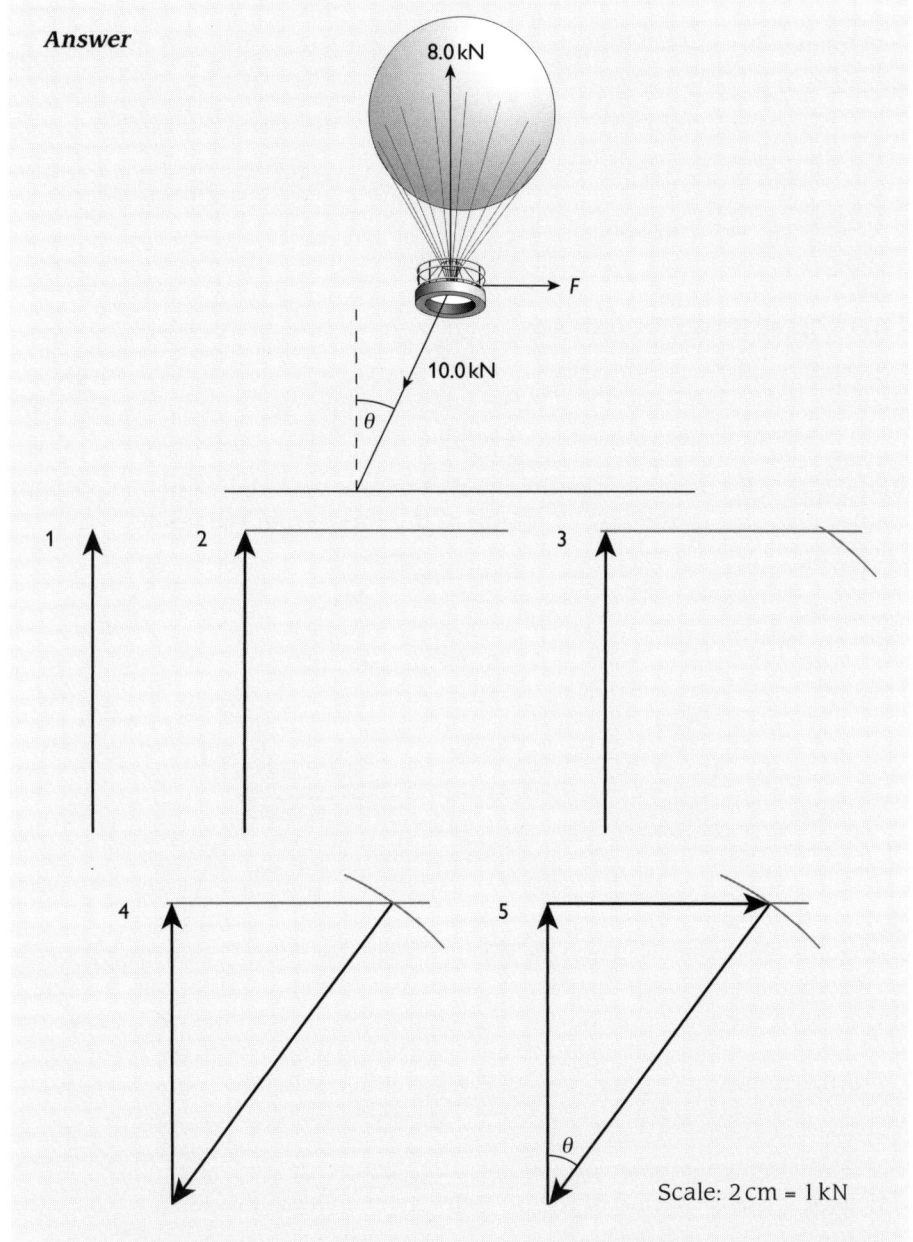

1 Draw a vertical arrow of length 4.0 cm to represent the upthrust on the balloon.

2 Draw a horizontal construction line from the top of the vertical arrow.

3 Separate the needle and the pencil tip of your compasses by a distance of 5 cm to represent the tension force in the cable. Place the needle on the bottom of the vertical line and draw an arc to intersect with the horizontal construction line.

    4 Join from the intersection to the bottom of the vertical line. This represents the tension in the cable.

    5 Draw in the arrow to represent the horizontal force from the wind.

    The length of the horizontal arrow = 3.0 cm. Therefore the force due to the wind is 6.0 kN. The angle with the vertical, θ, measured with a protractor = 37°.

Turning effects of forces

The turning effect of a force about a point (sometimes known as the **torque**) is known as its moment about that point. When considering a single force, the point about which the force is producing its turning effect must be specified.

Moment of a force about a point equals the force multiplied by the perpendicular distance of the line of action of the force from the point.

Consider a spanner being used to turn a nut:

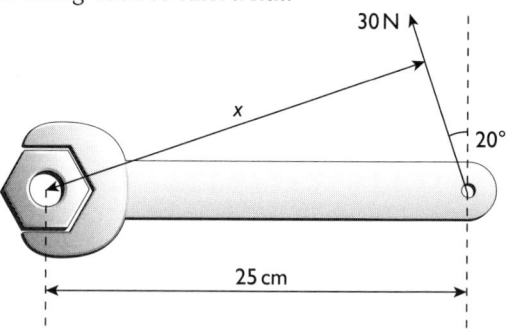

The force is not perpendicular to the spanner. Therefore either the component of the force perpendicular to the spanner or the perpendicular distance from the centre of the nut to the line of action of the force must be used in the calculation. The perpendicular distance of the line of action of the 30 N force from the centre of the nut is the distance $x = 25 \cos 20$ cm. Hence the torque about the centre of the nut is $30 \times 25 \cos 20 \approx 705$ N cm.

It is worth noting that torque is a vector. If a torque that tends to turn a body in a clockwise sense is considered to be positive, then a torque that tends to cause the body to move in an anticlockwise sense is considered negative.

Torque of a couple

A couple is produced when two parallel forces of equal magnitude act in opposite directions and have lines of action that do not coincide. You apply a couple when you turn on a tap. When considering the torque of a couple you do not need to worry about the specific point that the torque is produced about. The torque is the same whatever point is chosen. The torque of the couple is equal to the sum of the moments about any point of each of the two forces.

torque of a couple = magnitude of one of the forces × perpendicular distance between the lines of action of the forces

A couple tends to produce rotation only. A single force may tend to produce rotation but it will always tend to produce acceleration as well.

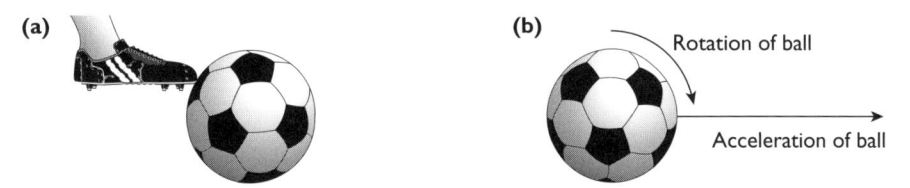

(a) A footballer kicks the ball striking the side of the ball
(b) The ball accelerates but also tends to rotate in a clockwise direction

Equilibrium

If the resultant force acting on a point object is zero then it is in equilibrium.

If, however, the body is of finite size then the possibility of rotational as well as translational movement must be considered.

For a body of finite size to be in equilibrium:
- the resultant force on the body must be zero
- the resultant torque on the body must be zero

Centre of gravity

The weight of a body does not act from a single point but is spread through all the particles of the body. However, it is often convenient to consider the weight acting at a single point — this point is called the **centre of gravity** of the body. A common term used in examinations is '**a uniform body**'. This means that the centre of gravity of the body is at the geometric centre of the body.

Principle of moments

The principle of moments is a restatement of the second condition for a body to be in equilibrium: **for a body to be in equilibrium, the sum of the moments about any point is zero.**

A useful way of using this when you are considering coplanar forces is to say 'the clockwise moments = the anticlockwise moments'.

> **Worked example**
> A student has a uniform metre ruler of weight 1.20 N. He attaches a weight of 1.50 N at the 10.0 cm point and places the ruler on a knife edge. He adjusts the knife edge until the ruler balances. Deduce the position of the knife edge.

Answer
Draw a diagram of the set up:

|──────────── 50 cm ────────────|
|────── x ──────|
|─ 10 cm ─|

1.50 N   1.20 N

The ruler is uniform and therefore the centre of gravity is at its centre.

Take moments about the pivot:
clockwise moment = $(50 - x) \times 1.20$ N cm
anticlockwise moment = $(x - 10) \times 1.50$ N cm

For equilibrium: the clockwise moments = the anticlockwise moments
$(50 - x) \times 1.2 = (x - 10) \times 1.5$
$x = 27.8$ cm

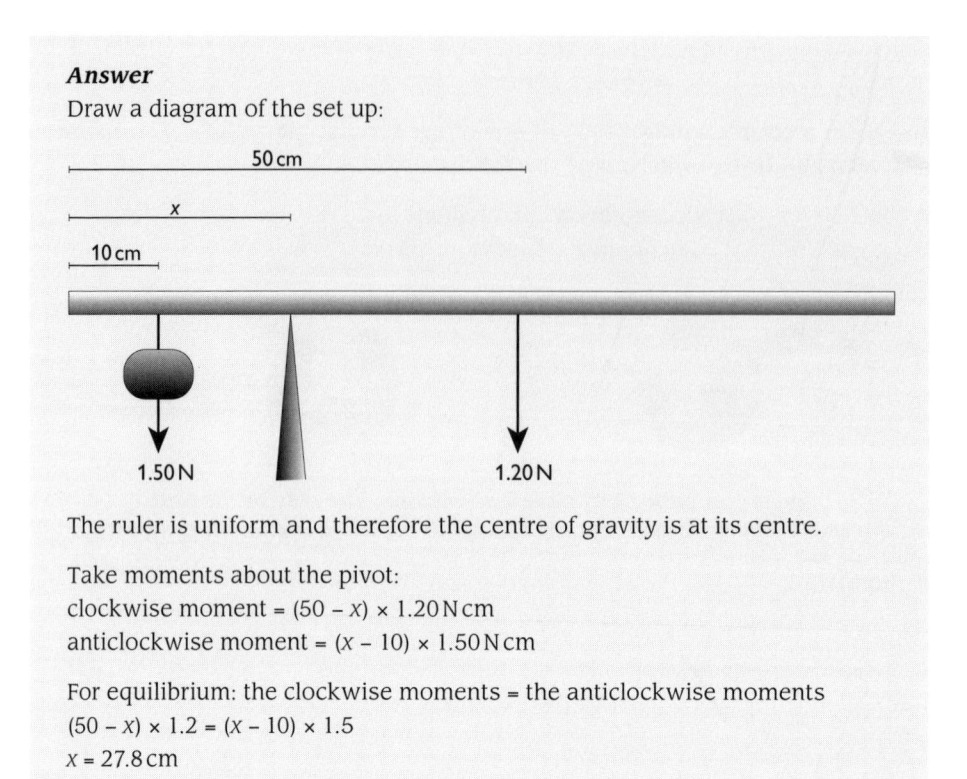

Work, energy and power

Work

Work has a precise meaning in physics and care must be taken when using this term. Work is defined as being done when a force moves its point of application in the direction in which the force acts:

work done = force × displacement in the direction of the force

The unit of work is the **joule (J)**. 1 joule of work is done when a force of 1 newton moves its point of application 1 metre in the direction of the force.

Both force and displacement are vectors. Note that for work to be done there must be a component of the force which is parallel to the displacement.

When calculating work done, care must be taken that the force and the displacement are parallel. Consider a child sliding down a slide.

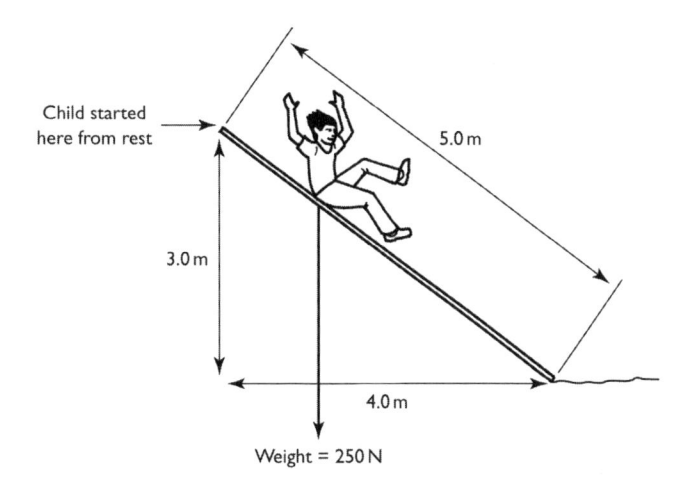

International AS and A Level Physics Revision Guide

The force on the child is the child's weight, 250 N, which acts vertically downwards. The total distance moved is 5.0 m but the displacement **parallel to the force** is only 3.0 m. So:

work done = 250 N × 3.0 m = 750 J

It is worth noting that in this example the work is done on the child by gravity, rather than the child doing work.

In general:

work done = component of the force parallel to displacement × the displacement

work done = $Fx\cos\theta$

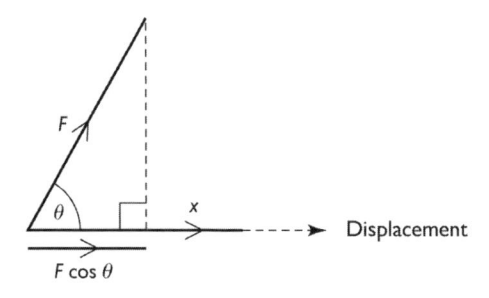

Worked example

The diagram shows a farmer wheeling a barrow.

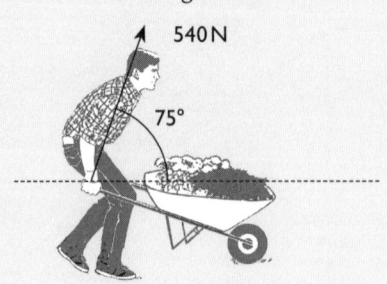

He applies a force of 540 N to the barrow in a direction 75° from the horizontal. He moves the barrow 30 m along the level ground. Calculate the work he does against friction.

Answer

work done = $Fx \cos \theta$

work done = 540 × 30 × cos 75 = 4200 J

Energy

Energy is not an easy concept and, like work, it has a precise meaning in physics. Energy is defined as the ability (or capacity) to do work. Energy is measured in joules. When a body has 300 J of energy it means that it can do 300 J of work.

Different forms of energy are shown in Table 3.

Table 3

Type of energy	Description
Kinetic energy	The ability to do work due to the movement of a body
Gravitational potential energy	The ability to do work due to the position of a body in a gravitational field
Elastic potential energy	The ability to do work due to the deformation of a body (e.g. a compressed or extended spring)
Sound energy	The ability to do work due to the kinetic and potential energy of the vibrating particles in a sound wave
Internal energy	The ability to do work due the random kinetic and potential energy of the particles in a body
Electrical potential energy	The ability to do work due to a the position of a charged particle in an electric field
Chemical potential energy	The ability to do work due to potential energy of the particles making up substances
Nuclear potential energy	The ability to do work due to the potential energy of the subatomic particles in the nuclei of atoms

Conservation of energy

The law of conservation of energy states that the total energy of a closed system is constant.

For examination purposes, you should explain this statement by saying that this means that energy can be transformed from one form to another but it can neither be created nor destroyed — the total energy of a closed system will be the same before an interaction as after it. When energy is transformed from one form to another either:
- work is done — for example, a man does work against gravity by lifting a large mass onto his shoulders.

or
- energy is radiated or received in the form of electromagnetic radiation — for example, internal energy is radiated away from the white hot filament of a lamp by infrared and light radiation.

Gravitational potential energy

Consider a mass m lifted through a height h.

The weight of the mass is mg, where g is the gravitational field strength.

work done = force × distance moved

$\quad = mg\Delta h$

Due to its new position, the body is now able to do extra work equal to $mg\Delta h$. It has gained extra potential energy, $\Delta W = mg\Delta h$

change in potential energy = $mg\Delta h$

If we consider a body to have zero potential energy when at ground level, we can say that:

gravitational potential energy (E_p) = mgh

In these examples we have considered objects close to Earth's surface, where we can consider the gravitational field to be uniform. In your A2 studies you will explore this further and consider examples where the gravitational field is not uniform.

Kinetic energy

Consider a body of mass m, at rest, which accelerates to a speed of v over a distance s.

$\quad$ work done in accelerating the body = force × distance

$\quad W = Fs$

but $F = ma$

In the equation $v^2 = u^2 + 2as$:

$\quad u = 0$

hence, $a = v^2/2s$

$$F = ma = \frac{mv^2}{2s}$$

$$W = Fs = m\frac{v^2}{2s}s = \tfrac{1}{2}mv^2$$

The body is now able to do extra work = $\tfrac{1}{2}mv^2$ due to its speed. It has **kinetic energy** = $\tfrac{1}{2}mv^2$

Worked example
A cricketer bowls a ball of mass 160 g at a speed of 120 km h^{-1}.

Calculate the kinetic energy of the ball.

Answer
Convert the speed from km h^{-1} to m s^{-1}:
120 km h^{-1} = 120 × (1000/3600) m s^{-1} = 33.3 m s^{-1}

Convert 160 g to kg = 0.16 kg.

$E_k = \tfrac{1}{2}mv^2 = \tfrac{1}{2} \times 0.16 \times 33.3^2$

$E_k = 89$ J

Strain energy
When a force causes an object to change its shape, the particles of the body are either squashed together or pulled apart. Therefore, they have extra potential energy. This is looked at in quantitative terms in the section on deformation of solids.

Internal energy
Internal energy is the sum of the random distribution of kinetic and potential energies of the particles of a system. When there is a rise in temperature, the average kinetic energy of the particles of a body (solid, liquid or gas) increases. When a material changes state from a solid to a liquid or from a liquid to a gas without a change in temperature, work is done in separating the particles giving them more potential energy. When they change from gas to liquid or liquid to solid the particles lose potential energy and energy is given out. Internal energy is explored further in the section 'Phases of matter'.

Energy losses and efficiency
Machines are used to do work, converting energy from one form to another. In practice, machines are never 100% efficient. This means that the total energy input is greater than the useful work output. Some of the energy input is converted to unwanted forms such as internal energy and sound.

efficiency of a machine = (useful work output/total energy input) × 100%

Efficiency is quoted either as a ratio or percentage. Consequently, efficiency has no units.

International AS and A Level Physics Revision Guide

Worked example
A petrol motor is used to lift a bag of sand of mass 2700 kg from the ground up to a window 12 m above the ground. Eighteen per cent of the input energy is converted into gravitational energy of the sand.

(a) Calculate the energy input to the motor.

(b) Discuss the energy changes involved in the process.

Answer

(a) efficiency = (useful work done/energy input) × 100%

useful work done = mgh = 2700 × 9.8 × 12 = 317 520 J

18 = (317 520/ energy input) × 100%

energy input = (317 520 × 100)/18 = 1 764 000 J ≈ 1.8 MJ

(b) The chemical potential energy of the petrol is converted into internal energy in the motor and 18% of this is used to do work against gravity in lifting the sand. The remainder is lost to internal energy as the surroundings are heated.

Tip

Note that no attempt has been made to discuss what happens inside the motor. This is extremely complex. There are transient forms of energy, such as strain potential energy, as the fuel is burnt and put under pressure, and the conversion of this to kinetic energy of the oscillating piston. Attempts to discuss what happens inside the motor are unlikely to succeed and should be avoided.

Power

Power (*P*) is the rate of doing work or transforming energy.

power = work done/time taken = energy transformed/time taken

Power is measured in joules per second. One joule per second is **1 watt**. The symbol for watt is **W**.

Worked example
A pebble of mass 120 g is fired from a catapult. The pebble accelerates from rest to 15 m s^{-1} in 0.14 s.

Calculate the average power gain of the pebble during the firing process.

53

Answer

120 g = 0.12 kg

gain in kinetic energy = $\frac{1}{2}mv^2$ = 0.5 × 0.12 × 15^2 = 13.5 J

power gain = 13.5/0.14 = 96 W

Power and velocity

Consider a car travelling at constant velocity v along a straight, level road. The engine must continue to do work against friction. If the frictional force is F, then the engine will supply an equal-sized force in the opposite direction. The work done by the engine, ΔW, in time Δt is $F\Delta s$, where Δs is the distance travelled in time Δt.

power = $F\Delta s/\Delta t$

but $\Delta s/\Delta t = v$, therefore:

power = Fv

Worked example

A cyclist is travelling along a straight level road at a constant velocity of 27 km h^{-1} against total frictional forces of 50 N.

Calculate the power developed by the cyclist.

Answer

Convert the velocity from km h^{-1} into m s^{-1}: 27 km h^{-1} = 27 × 1000/3600 = 7.5 m s^{-1}

power = force × velocity
= 7.5 × 50 = 375 W

Matter

Phases of matter

Density

Density is defined as the mass per unit volume:

density (ρ) = mass/volume

The unit of density is kilogram per metre cubed (kg m^{-3}) or gram per centimetre cubed (g cm^{-3}).

> **Worked example**
>
> A beaker has a mass of 48 g. When 120 cm³ of copper sulfate solution are poured into the beaker it is found to have a mass of 174 g. Calculate the density of the copper sulfate solution.
>
> *Answer*
>
> mass of copper sulfate solution = 174 − 48 = 126 g
>
> density = mass/volume = 126/120 = 1.05 g cm⁻³

Kinetic model of matter

Physics deals in models. The kinetic model of matter is a powerful model that can explain and predict macroscopic properties of materials. For example, the densities of solids are generally greater than the densities of liquids. This is because the particles in solids are closer together than the particles in liquids. The densities of gases are much less than the densities of solids and liquids. This is because the particles in gases are much further apart than those in liquids or solids.

The kinetic model is based on the following ideas:
- Matter is made up of small particles (atoms, ions or molecules).
- The particles move around.
- There are forces between the particles.

Solids, liquids and gases

The three phases of matter can be distinguished on a macroscopic level and on a microscopic level. Table 4 shows the differences.

Table 4

Phase	Macroscopic properties	Microscopic properties
Solid	Definite volume, definite shape	Particles are in fixed positions about which they can vibrate; the interparticle forces are large
Liquid	Definite volume, takes the shape of the container	Particles are further apart than in a solid and are free to move around the body of the liquid; the interparticle forces, although much smaller than in a solid, are still significant
Gas	Neither definite volume nor shape, completely fills a container	Particles are much further apart and can move around freely; interparticle forces are negligible

Solids can have different structures. Table 5 summarises the different type of solids.

Table 5

Type of solid	Structure	Examples
Crystalline	Highly ordered with the particles in fixed geometric patterns	Sodium chloride (common salt), diamond The geometric patterns of the ions or atoms determine the geometric shapes of the crystals
Metallic	Made up of many very small crystals (polycrystalline); the order is limited by the size of the crystals	Cast iron, copper The tiny crystals can be seen under a microscope when the material fractures; crystals in cast iron tend to be larger than those in copper and can sometimes be seen by the naked eye
Polymeric	Long-chain molecules	Rubber, polythene Rubber is a natural polymer; the chain molecules tend to be tangled and when rubber is stretched they straighten out, giving rubber its natural springiness Polythene is a man-made polymer; it is an example of a plastic
Amorphous	No long-term order in the molecular structure	Glass, wax Amorphous solids do not have a fixed melting point but gradually soften over a range of temperatures

Brownian motion

The first convincing evidence of the movement of particles came from a Scottish botanist, Robert Brown. He observed the random zigzag motion of pollen grains floating on water. Other people had observed this before but had assumed that it was the 'live' nature of pollen that caused the movement. Brown tried other materials, such as finely ground stone, and found the same behaviour.

The effect can be demonstrated in various ways. One method is to use a smoke cell, in which tiny globules of oil from the smoke are illuminated and viewed under a microscope. The diagram below shows a similar experiment in which the movement of tiny fat globules suspended in very dilute milk are studied.

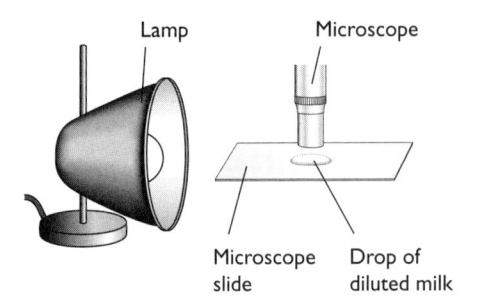

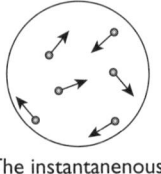

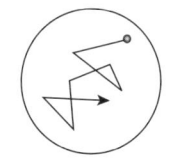

Tiny bright specks of light are seen through the microscope. These are caused by light being scattered by the fat globules. The specks of light move in random zigzag paths. The movement is due to the fat globules being bombarded by the much smaller water molecules. Although the water molecules themselves cannot be seen, their effects are seen. This experiment gives strong evidence in support of the existence of molecules and their movement.

We know that the water molecules are very much smaller than the fat globules because the movements, which are caused by unequal collisions on the different parts of the fat globules, are very small.

Pressure

Pressure is defined as the normal force per unit area:

pressure (*p*) = force/area

Pressure is measured in newtons per metre squared (N m^{-2}). 1 N m^{-2} is called 1 **pascal (Pa)**. It is sometimes convenient to use N cm^{-2}.

Pressure, unlike force, is a scalar. Therefore, pressure does not have a specific direction.

Worked example

Coins are produced by stamping blank discs with a die. The diameter of a blank disc is 2.2 cm and the pressure on the disc during stamping is 2.8×10^5 Pa. Calculate the force required to push the die against the blank disc.

Answer

area of the coin = $\pi (d/2)^2 = \pi (2.2/2)^2 = 3.8\,\text{cm}^2 = 3.8 \times 10^{-4}\,\text{m}^2$

pressure = force/area

hence, force = pressure × area = $2.8 \times 10^5 \times 3.8 \times 10^{-4} = 106\,\text{N}$

Pressure in a liquid

A liquid exerts pressure on the sides of its container and on any object in the liquid. The pressure exerted by the liquid increases as the depth increases. The diagram below shows a beaker containing a liquid of density, ρ.

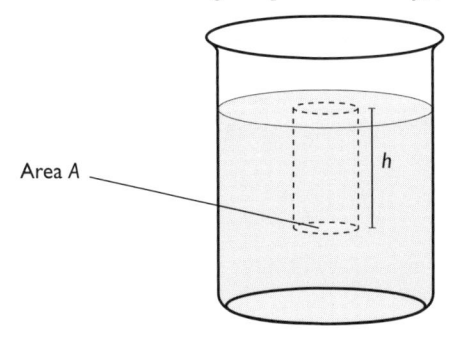

The pressure on the area, *A*, is due to the weight of the column of water of height, *h*, above it.

weight = mass × g (where g is the gravitational field strength)

mass of the column = density × volume, where the volume of the column of water = A × h

mass of the column = ρ × A × h

weight of the column = ρ × A × h × g

pressure on area, A, = force/area = weight/area

= ρ × A × h × g/A

pressure = ρhg

Worked example
Atmospheric pressure is 1.06×10^5 Pa. A diver descends to a depth of 24 m in sea water of density 1.03×10^3 kg m^{-3}. Calculate the total pressure on the diver.

Answer
pressure due to sea water = $h\rho g$ = $24 \times 1.03 \times 10^3 \times 9.8 = 2.42 \times 10^5$ Pa

total pressure = $2.42 \times 10^5 + 1.06 \times 10^5 = 3.48 \times 10^5$ Pa

Hint

Note that the total pressure is equal to the pressure due to the water plus atmospheric pressure.

You can now see clearly how upthrust (or buoyancy force) is produced. Consider a rectangular box in a liquid — the bottom of the box is at a greater depth than the top. Thus the pressure on the bottom is greater than the pressure on the top. Since the two surfaces have the same area, the force on the bottom is greater than the force on the top and the box is pushed upwards.

Pressure in gases
The movement of the molecules of a gas produces a pressure on the walls of its container.

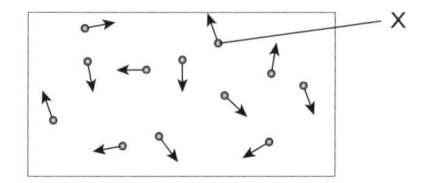

The diagram shows the random movement of molecules in a container. The molecule labelled X is about to hit the container wall. When it does, it will be a perfectly elastic collision and the molecule will rebound. Momentum is a vector quantity so the molecule undergoes a change in momentum. Hence the wall exerts a force on the molecule and the molecule exerts an equal-sized force on the wall in the opposite direction. (Newton's second and third laws). There will be millions of molecules in the container and they will continually bombard the walls, producing outward forces on the walls. The pressure on the walls is equal to the sum of all those forces divided by the area of the container walls.

Expansion of gases

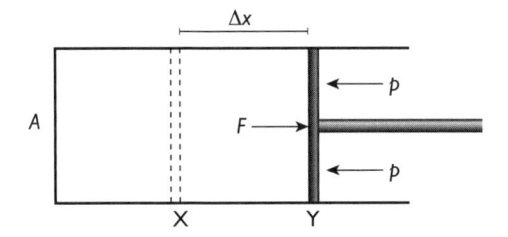

The diagram shows a gas contained in a cylinder of cross section A, with a friction-free piston. The gas is heated so that its pressure remains constant as it expands. The piston moves from X to Y against a constant external pressure p.

The piston does work W against the external pressure:

$W = F\Delta x$

$p = F/A$, therefore $F = pA$ and $W = pA\Delta x$

$A\Delta x$ = the change in volume of the cylinder ΔV, hence

$\Delta W = p\Delta V$

Change of state

When the temperature of a body is increased, the average random kinetic energy of the particles is increased. In gases and liquids, it is the average translational kinetic energy that increases; in solids, it is the average kinetic energy of vibration. When a pure material changes state, there is no change in temperature. The energy that is put in to change the state goes into increasing the potential energy of the particles, *not* their kinetic energy.

Melting

Melting is the change from the solid to the liquid state. When a solid melts, the separation of the particles increases by a small amount and work has to be done to pull the particles apart against the large attractive forces. The energy supplied does this work and the potential energy of the particles is increased. However, their kinetic energy remains unchanged. Hence, the temperature remains unchanged until all the solid has melted.

Boiling

Boiling occurs when the temperature of a liquid reaches a stage where bubbles of vapour can form in the body of the liquid. Once boiling starts, the temperature of the liquid remains constant. Any further energy supplied goes to separating the particles against the remaining attractive forces. The kinetic energy of the particles remains unchanged and their potential energy increases.

Evaporation

In both evaporation and boiling, liquid is turning into vapour (or gas). However, the two processes are quite different. Evaporation occurs at a range of temperatures; boiling occurs at a single fixed temperature. Bubbles of vapour form in the body of the liquid in boiling; in evaporation, particles leave the surface of the liquid.

If there is no external source of energy, the temperature of an evaporating liquid falls. Particles in any substance have a range of different energies. Those most likely to escape from a liquid are the particles with the greatest kinetic energy. When a liquid evaporates, the most energetic particles escape, so the average kinetic energy of the particles left behind is reduced. Hence, the temperature is lower.

Deformation of solids

We have already seen how forces produce changes in the motion of bodies; they can also change the shape of bodies. Forces in opposite directions will tend to stretch or compress a body. If two forces tend to stretch a body they are described as **tensile**. If they tend to compress a body they are known as **compressive**.

Forces on a spring

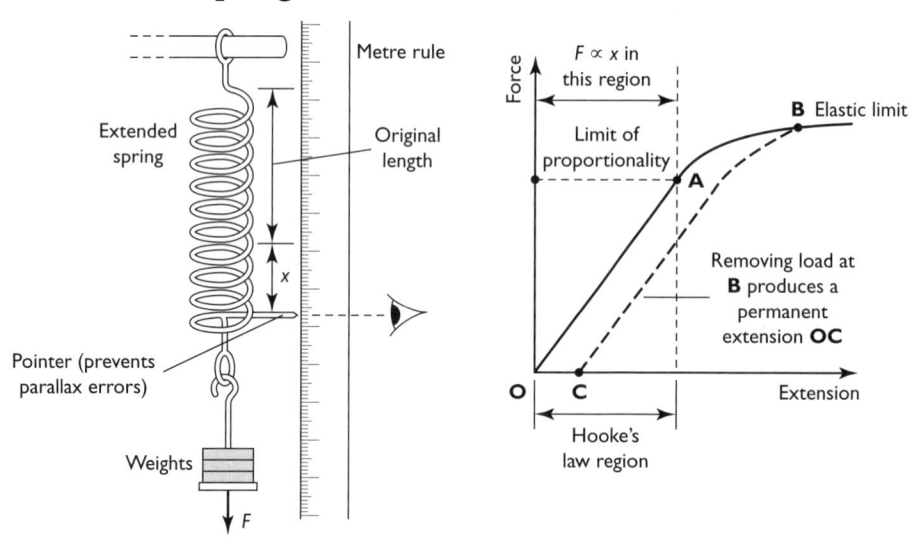

The diagram shows apparatus used to investigate the extension of a spring under a tensile force. The graph shows the results of the experiment. Analysing the results we see:
- From O to A the extension of the spring is proportional to the applied force.
- With larger forces, from A to B, the spring extends more easily and the extension is no longer proportional to the load.

- When the force is reduced, with the spring having been stretched beyond point B, it no longer goes back to its original length.

From O to A, F is proportional to x: $F \propto x$

This can be written as an equality by introducing a constant of proportionality:

$$F = kx$$

where k is the constant of proportionality, often known as the **spring constant**. The spring constant is the force per unit extension. It is a measure of the stiffness of the spring. The larger the spring constant, the larger is the force required to stretch the spring through a given extension. The unit of the spring constant is **newton per metre** (Nm^{-1}).

Point A, the point at which the spring ceases to show proportionality, is called the limit of proportionality. Very close to this point, there is a point B called the **elastic limit**. Up to the elastic limit, the deformation of the spring is said to be **elastic**. This means that the spring will return to its original length when the load is removed. If the spring is stretched beyond the elastic limit it will not return to its original length when the load is removed. Its deformation is said to be **plastic**.

Hooke's law sums up the behaviour of many materials that behave in a similar manner to a spring:

The extension of a body is directly proportional to the applied force.

Note that Hooke's law also applies to the compression of a body; in this case the quantity x in the equation is the compression rather than the extension.

Extension of a wire

The diagram below shows the apparatus that could be used to investigate the stretching of a wire.

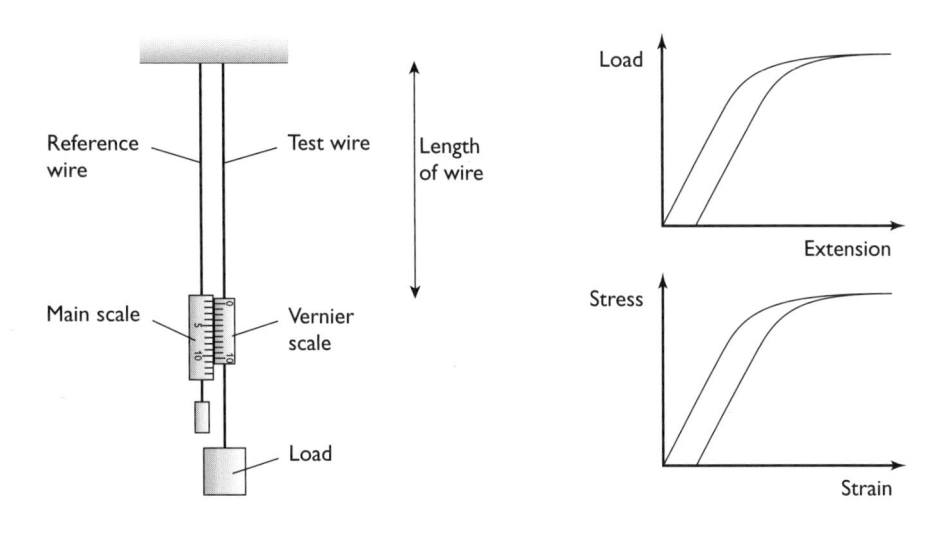

The readings that need to be taken are shown in Table 6.

Table 6

Reading	Reason	Instrument
Length of wire	Direct use	Metre ruler
Diameter of wire	Enables the cross-sectional area to be found	Micrometer screw gauge
Initial and final readings from the vernier slide	The difference between the two readings gives the extension	Vernier scale

The graph obtained is similar to that obtained for the spring. This shows the general nature of Hooke's law.

It is useful to draw the stress–strain graph which gives general information about a particular material, rather than for a particular wire.

Stress is defined as the force per unit cross-sectional area of the wire. The unit of stress is newton per metre squared or pascal (N m^{-2} = Pa).

Strain is the extension per unit of the unloaded length of the wire. This is a ratio and does not have units.

The formal symbol for stress is σ (the Greek letter sigma).

The formal symbol for strain is ε (the Greek letter epsilon).

The quantity stress/strain gives information about the elasticity of a material. This quantity is called the **Young modulus.**

Young modulus = stress/strain

= (force/area)/(extension/length)

= (force × length)/(area × extension) = $\dfrac{FL}{Ax}$

The unit of the Young modulus is the same as for stress, the **pascal** (Pa).

Worked example

A force of 250 N is applied to a steel wire of length 1.5 m and diameter 0.60 mm. Calculate the extension of the wire.

(*Young modulus for mild steel* = 2.1 × 10^{11} *Pa*)

Answer

Cross-sectional area of the wire = $\pi(d/2)^2$ = $\pi(0.6 \times 10^{-3}/2)^2$ = 2.83 × 10^{-7} m^2

Young modulus = $\dfrac{FL}{Ax}$

$2.1 \times 10^{11} = \dfrac{250 \times 1.5}{2.83 \times 10^{-7} \times \Delta L}$

$\Delta L = \dfrac{250 \times 1.5}{2.83 \times 10^{-7} \times 2.1 \times 10^{11}}$ = 6.3 × 10^{-3} m = 6.3 mm

Energy stored in a deformed material

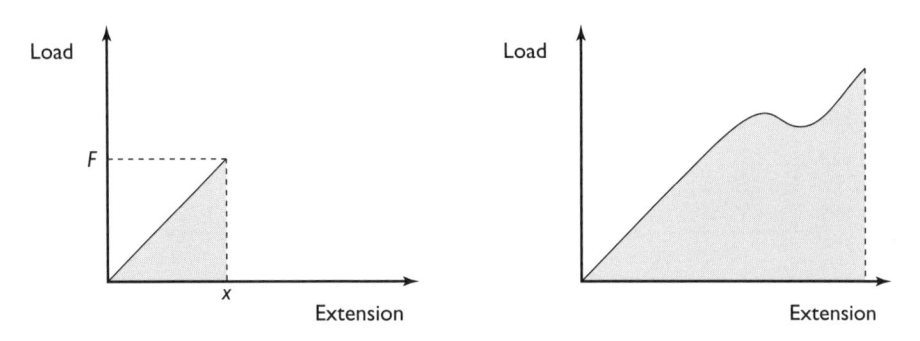

The first graph shows the extension of a body that obeys Hooke's law. The work done in stretching the body is equal to force multiplied by distance moved. This is equal to the strain potential energy in the body. The force is not, however, F the maximum force — it is the average force, which is ½F.

strain energy = ½Fx

This is the area of the triangle under the graph.

The general rule, even when the extension is not proportional to the load, is:

strain energy = area under the load–extension graph

It is worth noting that by substituting for F (= kx) in the original equation, it can be rewritten as:

strain energy = ½kx^2

This form of the equation shows us that the energy stored in the spring is proportional to the square of the extension — rather than the just the extension itself. This means that if the extension is doubled the energy stored is quadrupled, if the extension is tripled the energy stored is multiplied by nine etc.

If you consider the load–extension graph on page 61, you can see that the area under the graph when loading is larger than when unloading the wire. This means that more energy is stored in the stretched wire than is released when the load is removed. What happens to this energy? It is converted to internal energy in the wire — the temperature of the wire increases. The energy released is equal to the area in the enclosed loop made by the two curves. This is known as elastic hysteresis.

Different materials

Different materials deform differently under stress. The graphs below show some examples.

(a) Brittle material

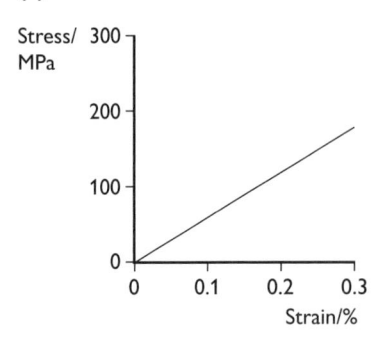

(b) Ductile material

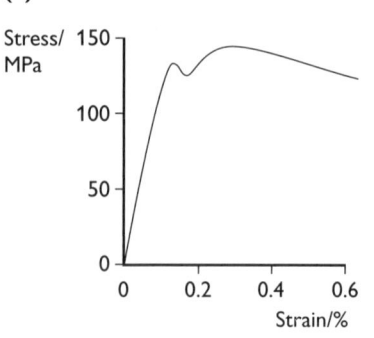

(c) Polymeric material

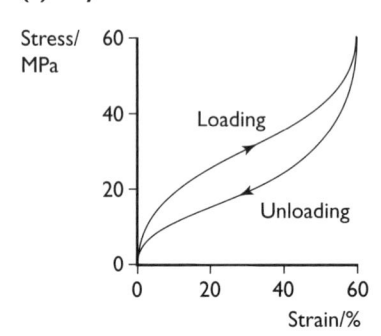

Behaviour of materials under stress (a) a brittle material (b) a ductile material and (c) a polymeric material

Brittle materials break at their elastic limit, with very little plastic deformation. Glass is a brittle material. It is surprisingly strong and has an **ultimate tensile stress** (the maximum tensile (stretching) stress that a material can withstand before fracture) of about 150 MPa. However, this varies widely because any small cracks on the surface will rapidly widen and reduce its strength.

Ductile materials initially stretch elastically obeying Hooke's law. However, once they reach their elastic limit they stretch much more per unit increase in load. They reach a point at which they continue to stretch, even if the load is reduced slightly. They then break. Most pure metals are ductile. Copper is a good example. It has about the same ultimate tensile stress as glass (150 MPa) but it stretches much more.

Polymeric materials are highly temperature dependent. At low temperatures they can act as brittle materials. At higher temperatures, their behaviour changes dramatically. Note that there is no standard shape of graph for polymerics; different polymers behave in different ways. Diagram (c) above shows the graph for rubber. Note how, once again, there is more work done in stretching the rubber than is released when it returns to its original length. The extra energy is released as internal energy in exactly the same way as when a metal wire is stretched beyond its elastic limit.

Oscillations and waves

Waves

In this course you will meet various types of wave. The importance of waves is that they are a way of storing energy (**stationary waves**) and transferring energy from one place to another (**progressive waves**).

Terminology

In mechanical waves, particles oscillate about fixed points. When a wave passes along a rope the particles of the rope vibrate at right angles to the direction of transfer of energy of the wave. Water waves can also be considered to behave in a similar manner. This type of wave is called a **transverse wave** (see diagram (a) below). Sound waves are rather different. The particles vibrate back and forth parallel to the direction of transfer of energy of the wave. This forms areas where the particles are compressed together (**compressions**) and areas where they are spaced further apart than normal (**rarefactions**). This type of wave is called a **longitudinal wave** (diagram (b) below).

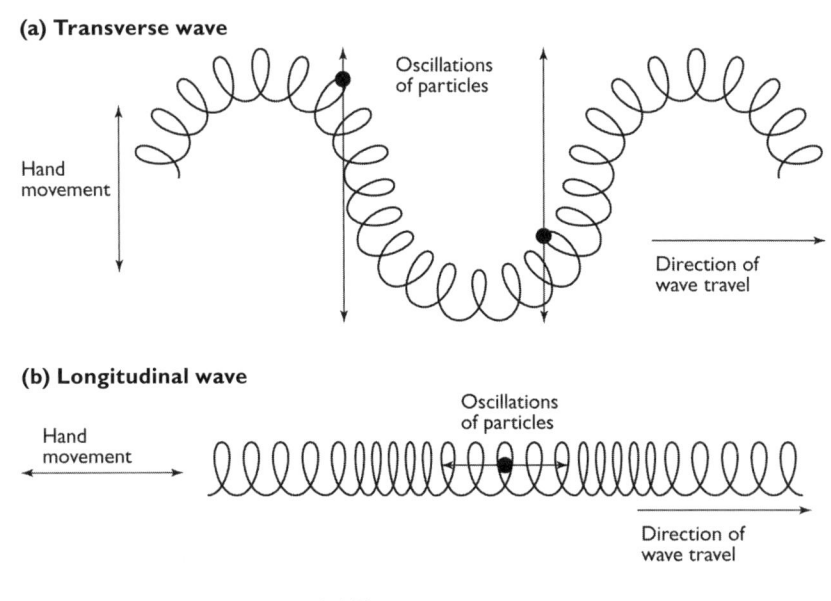

(a) Transverse wave
(b) Longitudinal wave

The graphs below show the displacement of a particle in a wave against time and the displacement of all the particles at a particular moment in time.

IV Oscillations and waves

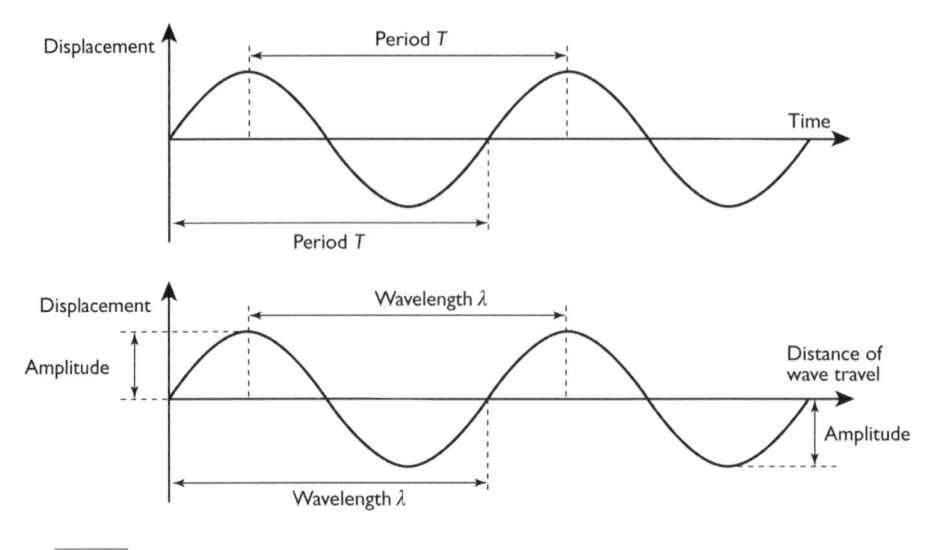

Hint

The shapes of these two graphs are similar. You should check the axes carefully to be sure of what the graph represents.

- **Displacement** (x) of a particle is its distance from its rest position. The unit is the metre (m).
- **Amplitude** (x_o) is the maximum displacement of a particle from its equilibrium position. The unit is the metre (m).
- **Period** (T) is the time taken for one complete oscillation of a particle in the wave. The unit is the second (s).
- **Frequency** (f) of a wave is the number of complete oscillations of a particle in the wave per unit time. The unit is the hertz (Hz).
- **Wavelength** (λ) is the distance between points on successive oscillations of the wave that are vibrating exactly in phase. The unit is the metre (m).
- **Wave speed** (c) is the distance travelled by the wave energy per unit time. The unit is the metre per second (m s^{-1})

1 hertz is defined as 1 complete oscillation per second.

An oscillation is one vibration of a particle — for example, from its rest position to the position of maximum displacement in one direction, back to the rest position, then to maximum displacement in the opposite direction and finally back to the rest position.

Frequency and period are related by the equation:

$f = 1/T$

The wave equation

The speed of a particle is given by the equation:

speed = distance/time

Similarly:

> wave speed = distance travelled by the wave/time

In time T, the period of oscillation, the wave travels one wavelength. Hence:

$c = \lambda/T$

but $T = 1/f$, hence, **$c = \lambda f$**

> **Worked example**
>
> A car horn produces a note of frequency 280 Hz. Sound travels at a speed of 320 m s^{-1}. Calculate the wavelength of the sound.
>
> **Answer**
>
> $c = f\lambda$
>
> $320 = 280\lambda$
>
> Therefore $\lambda = 320/280 = 1.14 \approx 1.1$ m

Phase difference

Moving along a progressive wave, the vibrating particles are slightly out of step with each other — there is a **phase difference** between them.

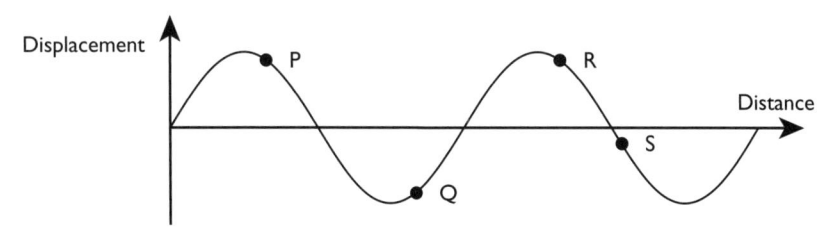

Study Table 7 which describes the phase relationships between the different points on the wave in the diagram.

Table 7

Points	Phase difference/ degrees	Phase difference/ radians	Common terms used to describe the phase difference
P and R	360 or 0	2π or 0	In phase
P and Q	180	π	Exactly out of phase (antiphase)
R and S	90	$\tfrac{1}{2}\pi$	90° or $\tfrac{1}{2}\pi$ out of phase

Phase difference also describes how two sets of waves compare with each other. The graphs below show two sets of waves that are approximately 45° ($\tfrac{1}{4}\pi$) out of phase. Phase difference is measured in degrees in AS work. You will meet radian measurements in the A2 section (page 132).

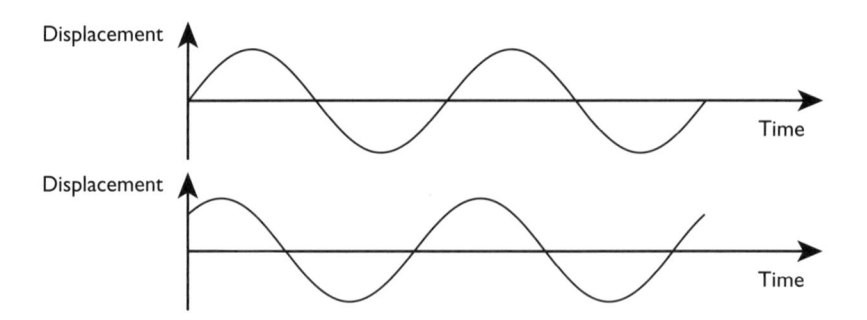

Transfer of energy

Progressive waves transfer energy. This can be seen with waves on the sea — energy is picked up from the wind on one side of an ocean and is carried across the ocean and dispersed on the other side, as the wave crashes onto a shore.

Electromagnetic waves

You have met the idea of energy being transferred by giving out and receiving radiation. This radiation consists of electromagnetic waves. The waves described previously are caused by the vibration of atoms or molecules. Electromagnetic waves are quite different — they are produced by the repeated variations in electric and magnetic fields. Electromagnetic waves have the amazing property of being able to travel through a vacuum. You see light (a form of electromagnetic wave) that has travelled through billions of kilometres of empty space from distant stars.

Electromagnetic radiation comes at many different frequencies. Table 8 lists different types of electromagnetic radiation and their approximate wavelengths in a vacuum.

Table 8

Type of radiation	Approximate range of wavelength in a vacuum/m	Properties and uses
Gamma radiation	10^{-16} to 10^{-11}	Produced by the disintegration of atomic nuclei; very penetrating, causes ionisation, affects living tissue
X-radiation	10^{-13} to 10^{-9}	Produced from rapidly decelerated electrons; properties similar to gamma-rays, the only real difference is in their method of production
Ultraviolet	10^{-9} to 4×10^{-7}	Ionising radiation, affects living tissue, stimulates the production of vitamin D in animals
Visible light	4×10^{-7} to 7×10^{-7}	Stimulates light-sensitive cells on the retina of the human (and other animals) eye
Infrared	7×10^{-7} to 10^{-3}	Has a heating effect and is used for heating homes and cooking
Microwaves	10^{-3} to 10^{-1}	Used in microwave cooking where it causes water molecules to resonate; also used in telecommunications, including mobile telephones
Radio waves	10^{-1} to 10^{5}	Used in telecommunications

It is important to recognise that there are not sharp boundaries between these types of radiation. The properties gradually change as the wavelength changes. For example, it is not possible to give a precise wavelength at which radiation is no longer ultraviolet and becomes X-radiation.

One property that these radiations have in common is that they all travel at the same speed in a vacuum, a speed of $3.0 \times 10^8 \, \text{m s}^{-1}$. Consequently, if we know a radiation's frequency, we can calculate its wavelength in a vacuum.

> **Worked example**
> The shortest wavelength that the average human eye can detect is approximately 4×10^{-7} m, which lies at the violet end of the spectrum. Calculate the frequency of this light.
>
> **Answer**
> $c = f\lambda$
>
> therefore $f = c/\lambda$
>
> therefore $f = 3.0 \times 10^8 / 4 \times 10^{-7} = 7.5 \times 10^{14}$ Hz

Intensity of radiation in a wave

Intensity is defined as the energy transmitted per unit time per unit area at right angles to the wave velocity. Energy transmitted per unit time is the power transmitted, so that

  intensity = power/area

The unit is watts per metre squared (W m^{-2}).

The intensity of a wave is proportional to the amplitude squared of the wave:

  $I \propto x_0^2$

This means that if the amplitude is halved the intensity is decreased by a factor of 2^2.

> **Worked example**
> The intensity of light from a small lamp is inversely proportional to the square of the distance of the observer from the lamp, that is $I \propto 1/r^2$. Observer A is 1.0 m from the lamp; observer B is 4.0 m from the lamp. Calculate how the amplitude of the light waves received by the two observers compares.
>
> **Answer**
> intensity of the light at B is $1/4^2 = 1/16$ of that at A
>
> intensity $\propto$ amplitude², therefore:
>
> amplitude $\propto \sqrt{\text{intensity}}$

amplitude at B = $1/\sqrt{16}$ that at A

amplitude at B = ¼ that at A

Hint

There are no simple formulae that you can apply here. You need to ensure that you understand the physics and then work through in a logical fashion.

Polarisation

In general, the oscillations in an electromagnetic wave do not just vibrate in a single plane. They vibrate in all directions perpendicular to the direction of travel of the wave energy.

(a)

Unpolarised light represented in three dimensions

(b)

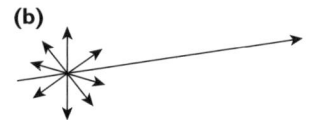

Vector diagram showing unpolarised light

(c)

Vector diagram showing the unpolarised light resolved into components at right angles

(d)

Vector diagram showing polarised light

Diagram (a) is a diagrammatic representation of such a wave. Diagram (b) shows a vector representation of this wave. These oscillations, like any vector, can be resolved at right angles to each other. This is shown in diagram (c). Diagram (d) shows the wave with the horizontal component filtered out. This type of wave is said to be **plane polarised**.

Light can be polarised by using a **polaroid** filter. This allows oscillations only in one direction to pass through. If a second polaroid filter, with its axis at right angles to the first is added, all the light is filtered out.

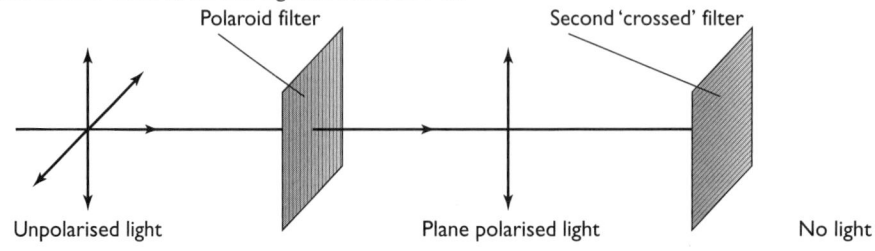

Unpolarised light → Polaroid filter → Plane polarised light → Second 'crossed' filter → No light

Radio waves and microwaves are usually plane polarised. This is why you can sometimes get a better signal if you turn a radio or telephone through different angles.

Identifying transverse waves
Longitudinal waves cannot be polarised because the vibrations are parallel to the direction of travel of the wave energy. Thus, if a wave can be polarised it means that it must be transverse.

Sound waves
Sound waves are longitudinal waves. Vibrations of the particles in this type of wave produce areas of pressure that are slightly higher (compressions) and slightly lower (rarefactions) than average. Consequently, sound waves are often described as being pressure waves.

Measurement of the frequency of a sound wave
The frequency of a sound wave can be measured using a cathode-ray oscilloscope.

The apparatus for this experiment is shown in the diagram below.

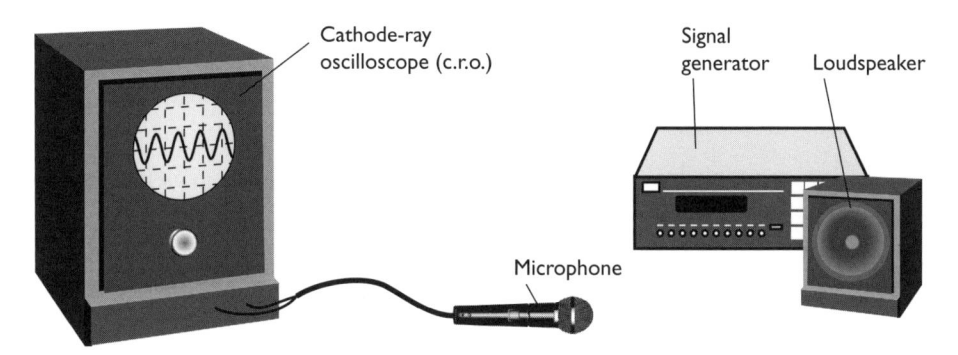

The period of the wave can be determined from the time-base setting and the number of waves shown on the screen. (frequency = 1/period)

> **Worked example**
> In the diagram above, the time base is set at $5\,\text{ms}\,\text{div}^{-1}$. Calculate the frequency of the wave.
>
> **Answer**
> In four divisions there are 3.5 waves.
>
> Therefore, in $4 \times 5\,\text{ms}$ (= 20 ms) there are 3.5 waves.
>
> Therefore, the period (time for 1 wave) = $20/3.5\,\text{ms} = 5.7 \times 10^{-3}\,\text{s}$
>
> $f = 1/T = 1/5.7 \times 10^{-3} = 175\,\text{Hz}$

> **Hint** Remember:
> - use as much of the screen as possible to reduce uncertainties
> - 1 wavelength is from one peak (or one trough) to the next peak (or trough)

Superposition

Stationary waves

If you pluck a stretched string at its centre it vibrates at a definite frequency, as shown in the diagram:

Vibration of string

This is an example of a stationary wave. It is produced by the initial wave travelling along the string and reflecting at the ends. It will die away because energy is lost to the surroundings, for example by hitting air molecules and producing a sound wave. This wave, where there is just a single loop, is called the **fundamental** wave or the **first harmonic**. Its wavelength is twice the length of the string. f_0

A more detailed description of how the wave is formed can be found on page 79.

A different stationary wave can be set up by plucking the string at points A and B. Note that the midpoint of the string has zero amplitude. This point is called a **node**. The points of maximum amplitude are called **antinodes**.

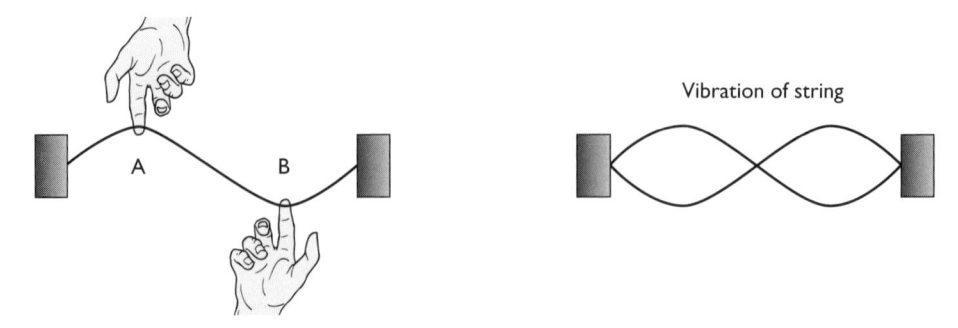

Vibration of string

The frequency of this wave is twice that of the previous wave and its wavelength is half that of the fundamental. It is called the second harmonic.

These waves die away quickly as energy is transferred to the surroundings. They can be kept going by feeding energy into the system:

Vibrator

A whole series of harmonics can be produced by varying the frequency of the vibrator. The first three are shown below:

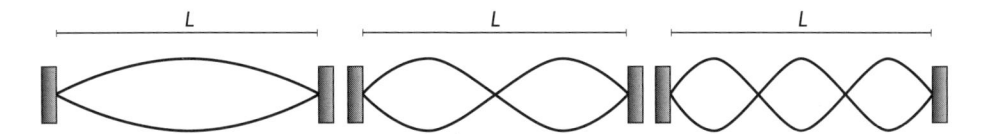

Note that each harmonic consists of a whole number of half wavelengths.

Hint

Remember that the distance between adjacent nodes, or between adjacent antinodes, is half a wavelength, *not* a full wavelength.

Stationary waves in air columns

Sound waves can produce stationary waves in air columns. Here, a small loudspeaker or a tuning fork is used to feed energy into the system.

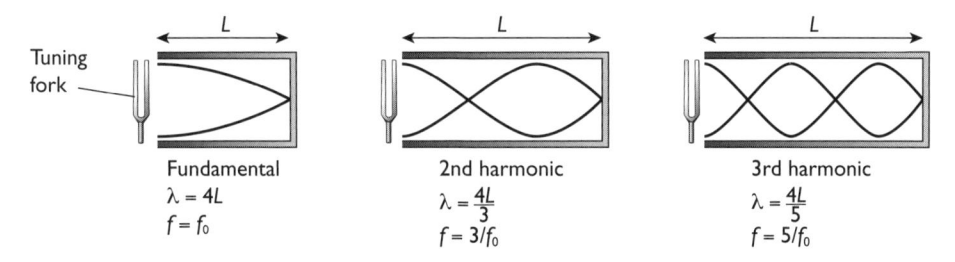

Fundamental
$\lambda = 4L$
$f = f_0$

2nd harmonic
$\lambda = \frac{4L}{3}$
$f = 3/f_0$

3rd harmonic
$\lambda = \frac{4L}{5}$
$f = 5/f_0$

Hint

The diagrams with sound waves are graphical representations showing displacement against position along the tube. Sound waves are longitudinal, so the displacement is parallel to the length of the tube *not*, as the diagrams can suggest, perpendicular to it.

Differences between stationary waves and progressive waves

Some differences between stationary and progressive waves are given in Table 9.

Table 9

Stationary waves	Progressive waves
Energy is stored in the vibrating particles	Energy is transferred from one place to another
All the points between successive nodes are in phase	All the points over one wavelength have different phases
The amplitudes of different points vary from a maximum to zero	All the points along the wave have the same amplitude

IV Oscillations and waves

Measurement of the wavelength of microwaves

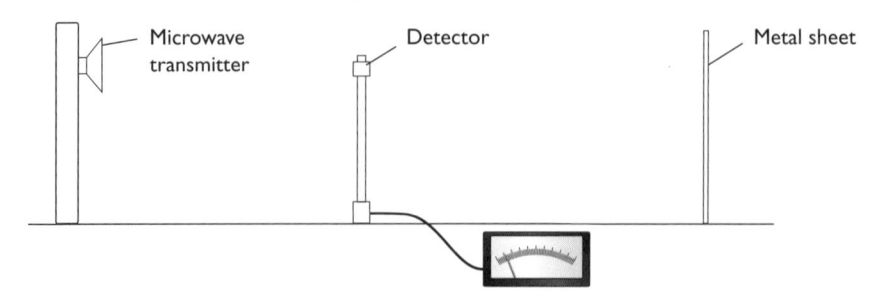

A stationary-wave pattern can be set up by reflecting microwaves from a metal sheet. As the detector is moved in a line from the transmitter and the metal sheet a series of maximum and minimum signals will be observed. This experiment, therefore, demonstrates stationary waves. As in previous examples, the wavelength of the microwaves is equal to the twice the distance between successive nodes or successive antinodes.

Measurement of the speed of sound

Apparatus for measuring the speed of sound is shown in the diagram below.

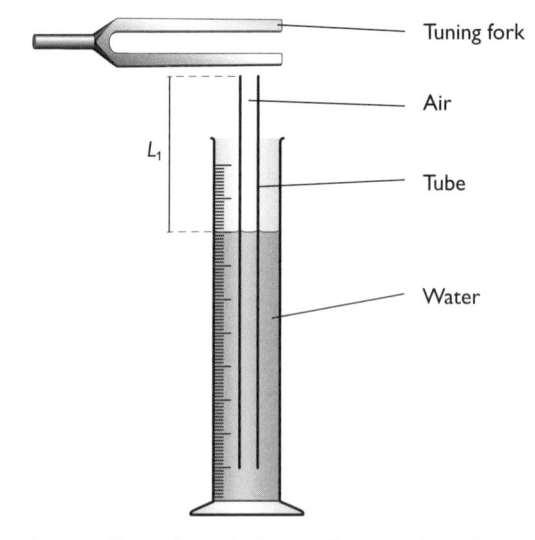

The height of the tube is adjusted until the fundamental stationary wave is formed. This can be identified by a clear increase in the loudness of the sound produced. The length L_1 is measured. The tube is then moved upwards until the next stationary wave is formed and the new length L_2 is measured. The wavelength is equal to $2(L_2 - L_1)$. If the frequency of the tuning fork is known then the speed of the sound in the air column can be calculated using the wave equation, $c = f\lambda$.

Hint

The antinode at the top of the tube is just beyond the top of the tube. This means that an end correction has to be allowed for. Subtracting the two readings eliminates the end correction.

Worked example
A tuning fork of frequency 288 Hz produces a stationary wave when a tube of air is 28.5 cm long. The length of the tube is gradually increased and the next stationary wave is formed when the tube is 84.0 cm long.

Calculate the speed of sound in the tube.

Answer

½λ = (84.0 − 28.5) = 55.5 cm

λ = 111 cm = 1.11 m

$c = f\lambda$ = 288 × 1.11 = 320 m s^{-1}

Diffraction

When waves pass through an aperture they tend to spread out. Similarly, if waves go round an object they tend to spread round it. The diagrams show **wavefronts** passing through a narrow and a wide slit and round an object.

A **wavefront** is a line on a wave which joins points which are exactly in phase.

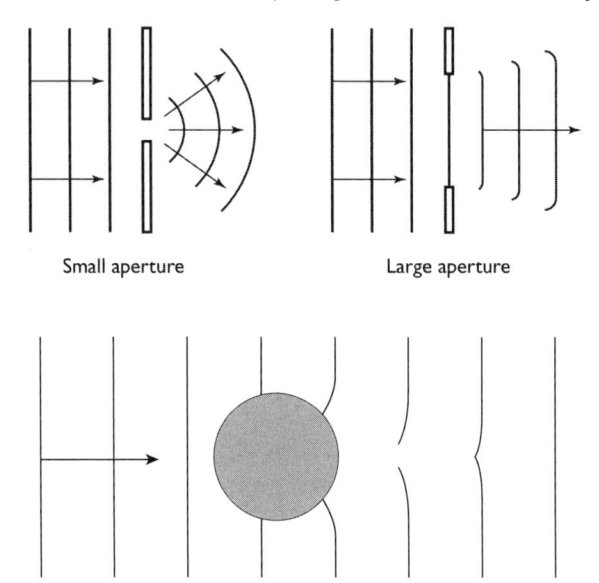

Small aperture   Large aperture

The greatest diffraction occurs when the aperture or object is the same diameter as the wavelength of the waves. If the object is much more than 100 times the size of the wavelength, then diffraction is much less noticeable. Visible light has a wavelength of about 10^{-6} to 10^{-7} m, so for diffraction to be observed the gap needs to be less than about 10^{-4} m (0.1 mm). You can observe diffraction of light by looking at a distant street light through a narrow gap made by your fingers.

IV Oscillations and waves

Interference

When two sets of waves of the same type meet, their displacements add or subtract in a similar way to vectors. At its most simple, if the two sets of waves are exactly in phase, the combined wave has an amplitude equal to the sum of the two amplitudes. This is known as **constructive interference**.

If the two sets of waves are 180° out of phase (in **antiphase**) the two waves will subtract. This is known as **destructive interference**.

If the original amplitudes are equal there will be no disturbance.

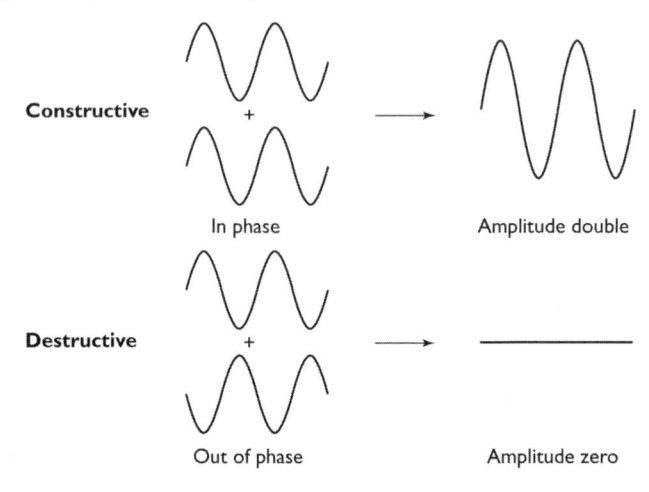

Constructive
In phase → Amplitude double

Destructive
Out of phase → Amplitude zero

Coherence

For interference to occur two **coherent** sources of waves are required. For the source to be coherent the waves must:
- be of the same frequency
- have a constant phase difference

Interference of sound

Interference of sound waves can be demonstrated using two loudspeakers driven by the same signal generator, thus giving coherent waves.

Loudspeakers

- Quiet
- Loud
- Quiet
- Point A
- Point B
- Point C
- Point D

- A loud sound is heard at A as waves from the two loudspeakers have travelled equal distances and are in phase. Therefore, the waves interfere constructively.

- A quiet sound is heard at B as waves from the upper loudspeaker have travelled half a wavelength further than waves from the lower speaker. Consequently, the waves are in antiphase, so they interfere destructively.
- A loud sound is heard at C as waves from the upper loudspeaker have travelled a full wavelength further than waves from the lower speaker. The waves are now in phase and so interfere constructively.
- A quiet sound is heard at D as waves from the upper loudspeaker have travelled one-and-a-half wavelengths further than waves from the lower speaker. The waves are now in antiphase, so interfere destructively.

Interference of water waves

Water waves can be shown interfering by using a ripple tank.

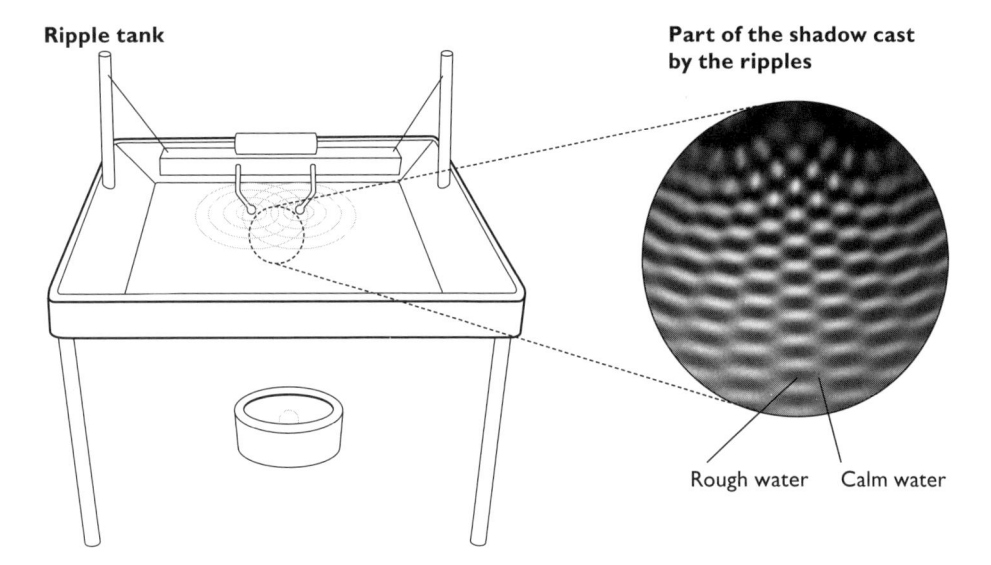

Ripple tank

Part of the shadow cast by the ripples

Rough water   Calm water

The areas of calm water (destructive interference) and rough water (constructive interference) can be viewed on the shadow image formed on the ceiling. Alternatively, they can be seen directly by looking almost parallel to the surface of the water.

Interference of light

Early attempts to demonstrate interference of light were doomed to failure because separate light sources were used. A lamp does not produce a continuous train of waves, it produces a series of short trains. The phase difference between one train and the next is random. Hence, if light from two separate sources is mixed there is no continuing relationship between the phases and an 'average brightness' is observed.

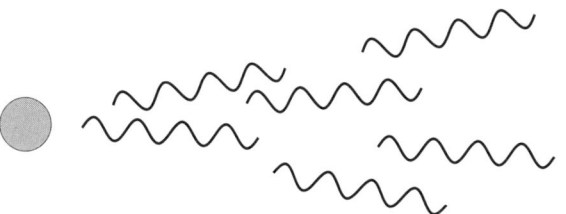

IV Oscillations and waves

To successfully demonstrate interference, the light from a single monochromatic source of light must be split and then recombined, with the two parts travelling slightly different distances.

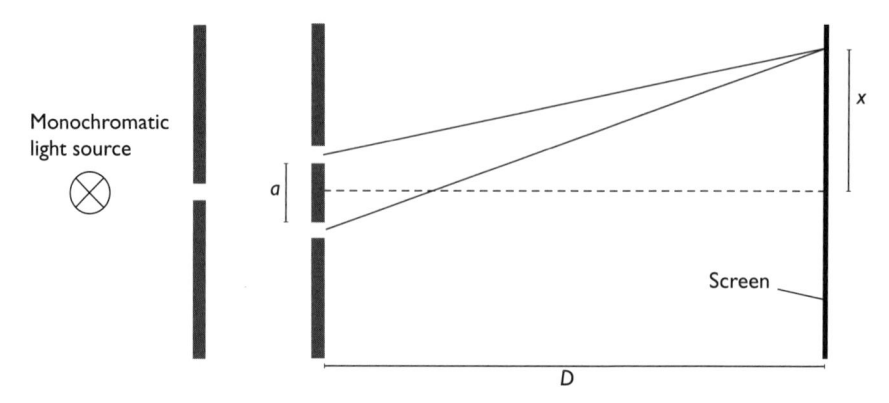

The wavelength of light is very short (~10^{-7} m). Consequently, the distance between the slits a must be small (<1 mm) and the distance D from the slits to the screen must be large (~1 m).

For constructive interference the path difference between the contributions from the two slits is $\frac{ax}{D}$, where x is the distance between adjacent bright fringes.

So, $\lambda = \frac{ax}{D}$

Worked example
Light of wavelength 590 nm is incident on a pair of narrow slits. An interference pattern is observed on a screen 1.5 m away. A student observes and measures 12 interference fringes, over a distance of 2.1 cm.

Calculate the separation of the two slits.

Answer

$\lambda = \frac{ax}{D}$

$x = 2.1/12$ cm $= 0.175$ cm $= 1.75 \times 10^{-3}$ m

$590 \times 10^{-9} = \frac{a \times 1.75 \times 10^{-3}}{1.5}$

$a = \frac{590 \times 10^{-9} \times 1.5}{1.75 \times 10^{-3}} = 5.1 \times 10^{-4}$ m

Note that the colour of light is dependent on its frequency. In general, for coherence we expect a single frequency, therefore single wavelength. In practice, when using white light (a whole range of colours) a few coloured fringes can be observed as the different wavelengths interfere constructively and destructively in different places.

Superposition and stationary waves

Formation of a stationary wave requires two waves of the same type and of the same frequency that travel in opposite directions and meet.

Consider a vibrating string producing a fundamental frequency stationary wave. The initial wave train travels along the string and is reflected at the far end. The reflected wave has the same frequency as the original wave. On the reflection, there is a 180° phase shift. Consequently, there is destructive interference at the point of reflection and the amplitude of the stationary wave is zero (a node). As we consider points on the string moving towards its centre, the phase difference between the incident and the reflected wave gradually decreases, so the amplitude of the stationary wave increases. At the centre of the string the two waves are in phase and the amplitude of the stationary wave is maximum (an antinode). As we go beyond the centre the waves gradually go further out of phase, until at the other end they are once more 180° out of phase and the amplitude is zero (a node).

For this to happen, the time that it takes the wave to travel from one end of the string to the other must be exactly equal to half the period of the wave. Consequently, the stationary wave is set up only when the string is made to vibrate at a specific frequency — the frequency it would vibrate at if it were 'plucked' at the centre. (This is known as the natural frequency of the string.)

The other harmonics are formed when the time taken for the wave to travel the length of the string is $\frac{1}{4}T$ (second harmonic), $\frac{1}{6}T$ (third harmonic) etc.

Multislit interference

The effect of using more than two slits to produce an interference pattern is to make the maxima sharper and brighter. The more slits there are, the sharper and brighter are the maxima. This makes it much easier to measure the distance between maxima.

Tip

The multislit device is called a diffraction grating, which is rather confusing. Although the spreading of the light (diffraction) is required for interference, this is really an interference grating!

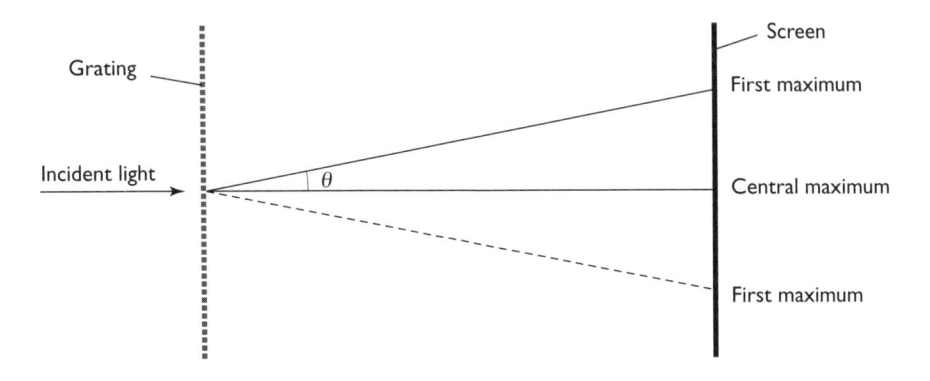

The path difference between contributions from successive slits is $d\sin\theta$, where d is the distance between successive slits. Hence for a maximum:

$$n\lambda = d\sin\theta$$

where n is a whole number. The first maximum ($n = 1$) is sometimes called the first order.

Worked example
Calculate the angles at which the first and second maxima are formed when a monochromatic light of wavelength 7.2×10^{-7} m is shone perpendicularly onto a grating with 5000 lines per cm.

Answer
$d = 1/5000\,\text{cm} = 2 \times 10^{-4}\,\text{cm} = 2 \times 10^{-6}\,\text{m}$

For the first maximum, $\lambda = d\sin\theta$
$\sin\theta = \lambda/d = 7.2 \times 10^{-7}/2 \times 10^{-6} = 0.36$
$\theta = 21°$

For the second maximum $2\lambda = d\sin\theta$
$\sin\theta = 2\lambda/d = 2 \times 7.2 \times 10^{-7}/2 \times 10^{-6} = 0.72$
$\theta = 46°$

Electricity and magnetism
Electric fields

An electric field is a region in which charged bodies experience a force. The **electric field strength** is defined as the force per unit positive charge on a stationary point charge.

electric field strength = force/charge, which may be written:

$$E = F/Q$$

The unit of electric field strength is newtons per coulomb (NC^{-1}).

We can represent the shape of a field by drawing lines of force. In an electric field the lines represent the direction of the force on a small positive test charge. When drawing an electric field:
- the direction of electric field lines is away from positive charges and towards negative charges
- the closer the field lines, the stronger the field strength
- the field lines never touch nor cross

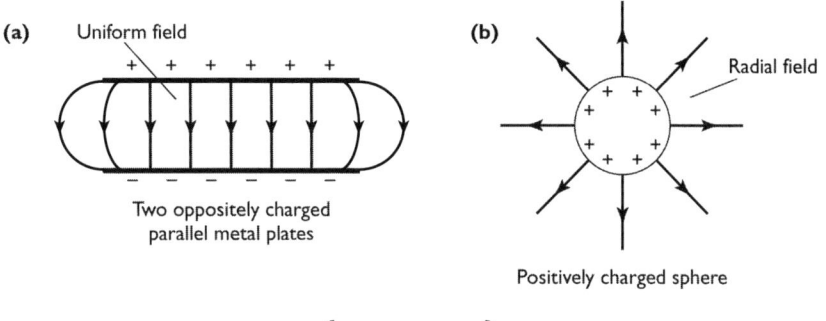

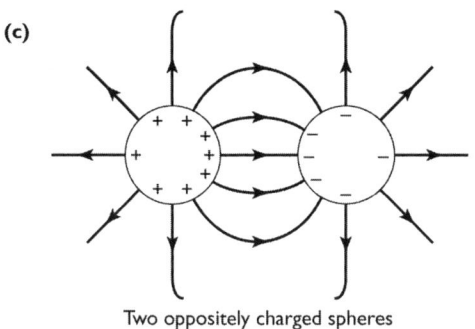

(a) Uniform field — Two oppositely charged parallel metal plates

(b) Radial field — Positively charged sphere

(c) Two oppositely charged spheres

The uniform electric field

You can see from diagram (a) above that once we get away from the edges of the plates, the field between two parallel plates is uniform. This means that wherever a charged particle is placed between those plates it experiences the same magnitude force, in the same direction.

The electric field strength between the plates is given by the formula:

$E = V/d$

where V is the potential difference and d is the distance between the plates. Note that this means that an alternative way of expressing the unit for electric field strength (NC^{-1}) is volts per metre (Vm^{-1}).

Worked example
A piece of dust carries a charge of -4.8×10^{-18} C, and lies at rest between two parallel plates separated by a distance of 1.5 cm. Calculate the force on the charge when a potential difference of 4500 V is applied across the plates.

Answer
$E = V/d = 4500/(1.5 \times 10^{-2}) = 300\,000\,Vm^{-1}$

$E = F/Q$

Therefore $F = EQ = 300\,000 \times (-4.8) \times 10^{-18} = -1.44 \times 10^{-12}\,N$

V Electricity and magnetism

Using electric fields

A charged particle in an electric field experiences a force and therefore tends to accelerate. If the particle is stationary or if the field is parallel to the motion of the particle, the magnitude of the velocity will change. An example is when electrons are accelerated from the cathode towards the anode in a cathode-ray tube.

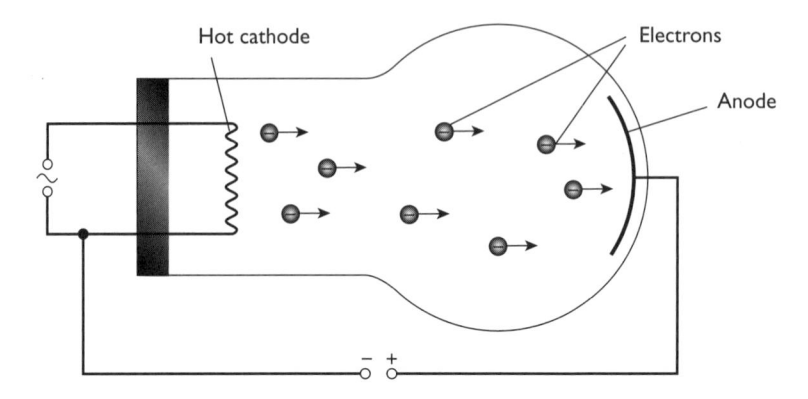

If the field is at right angles to the velocity of the charged particles, the direction of the motion of the particles will be changed. The path described by the charged particles will be parabolic, the same shape as a projectile in a uniform gravitational field. The component of the velocity perpendicular to the field is unchanged, the component parallel to the field increases uniformly.

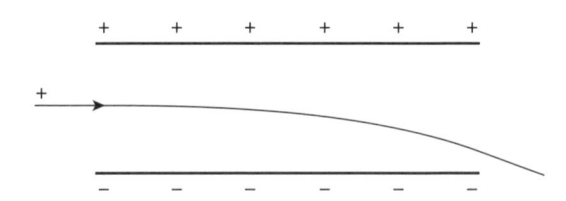

The path of a proton as it passes through a uniform electric field

The constant force on the charged particle leads to it describing a parabolic path. This path is similar to that of a ball thrown horizontally in a uniform gravitational field.

Current electricity

Electric current: terminology

It is important to be clear about the meanings of the different terms used in electricity.

Table 10

Quantity	Meaning	Unit and symbol
Current (*I*)	Movement of electric charge	**Ampere** (A)
Charge (*Q*)	'Bits' of electricity*	**Coulomb** (C)
Potential difference (*V*)	Work done in moving unit positive charge from one place to another place	**Volt** (V)
Resistance (*R*)	The opposition to current, it is defined as potential difference/current	**Ohm** (Ω)

*This is not a formal definition of charge. The concept of the nature of charge is quite complex. It can only be explained fully in terms of the interactions between charges and between charges and electric fields. However, we often think of charge in this way — consider the 'bit' (-1.6×10^{-19} C) which an electron carries.

A formal definition of charge is:

charge passing a point = current × time for which the current flows

Definitions of electrical units

Current

The **ampere** is one of the base units described in the opening section of this book. It is defined in terms of the force between two parallel conductors.

Charge

The **coulomb** is the charge which passes any point in a circuit when a current of 1 ampere flows for 1 second. This leads to equation 1:

$Q = It$

Potential difference

There is a potential difference of 1 **volt** between two points when 1 joule of work is done in moving a charge of 1 coulomb from one point to another. This leads to equation 2, where *W* is the work done in moving charge *Q*:

$V = W/Q$

Power

We met power in the section 'work, energy and power'. You should remember that:
- power is defined as the work done, or energy transferred, per unit time
- the unit is the watt (W), which is the power generated when work is done at the rate of 1 joule per second

This leads to equation 3:

$P = VI$

Resistance

A component has a resistance of 1 ohm (Ω) when there is a current of 1 ampere through the component and a potential difference of 1 volt across its ends. This leads to equation 4:

$R = V/I$

V Electricity and magnetism

Equations
1. $Q = It$
2. $V = W/Q$
3. $P = VI$
4. $R = V/I$

The following relationships can be found by substituting the resistance equation into the power equation:

$P = V^2/R$ and $P = I^2R$

Worked example
A water heater of resistance $60\,\Omega$ runs from a mains supply of 230 V. It can raise the temperature of a tank of water from 20 °C to 45 °C in 20 minutes.

Calculate:
(a) the charge that passes through the heater
(b) the energy dissipated by the heater

Answer
(a) $R = V/I$, which leads to $I = V/R = 230/60 = 3.83\,\text{A}$

$Q = It = 3.83 \times 20 \times 60 = 4600\,\text{C}$

(b) $V = W/Q$ which leads to $W = VQ = 230 \times 4600 = 1\,058\,000\,\text{J} \approx 1.1\,\text{MJ}$

Hint

Note that the information about the temperature rise of the water is irrelevant to the question. One skill you need to develop is to select relevant information and reject that which is not relevant.

I–V characteristics

Different components behave in different ways when a potential difference is put across them. Examples are shown in the diagrams below.

International AS and A Level Physics Revision Guide

Metal wire

Tungsten filament lamp

Thermistor

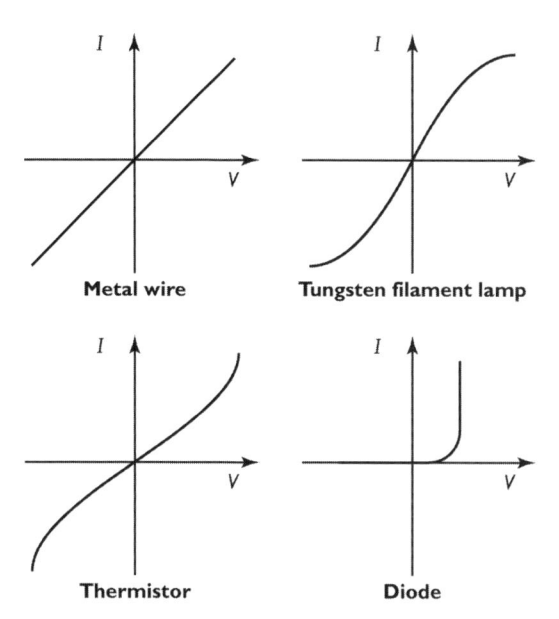

Diode

Table 11

Component	Description	Explanation
Metal wire	The current is proportional to the potential difference across it. The resistance is the same for all currents. The resistance of the wire is equal to the inverse of the gradient.	Metals contain many free electrons. These carry the current. The greater the potential difference, the greater the drift velocity of these electrons.
Filament lamp	At low currents, the current is proportional to the potential difference. At higher currents, the current does not increase as much for the same voltage increase. The resistance increases at higher currents.	Lamp filaments are made from tungsten metal. At low currents, the filament behaves in the same way as the wire. At higher currents, the temperature increases to around 1500°C, the vibrations of the ions in the crystal lattice increases, presenting a larger collision cross-section and reducing the drift velocity of the electrons.
Thermistor	At low currents, the current is proportional to the potential difference. At higher currents, the current increases more for the same voltage increase. The resistance decreases at higher currents.	Like the metal wire, for low currents the potential difference is roughly proportional to the current, but when the temperature increases the resistance decreases. Thermistors are semiconductors. Conduction in semiconductors is different from in metals. There are fewer free electrons. Increasing the temperature frees more electrons to carry the current and thus reduces the resistance.

V Electricity and magnetism

| Diode | No current will pass in one direction. Once the potential difference (in the opposite direction) reaches a set value (0.6 V for a silicon diode) it conducts with very little resistance. | Diodes are also semiconductors but they are designed to allow currents to pass in one direction only. |

Important note: for a filament lamp, a thermistor and a diode, the resistance of the component is *not* equal to the inverse of the gradient. It is equal to the potential difference divided by the current when that p.d. is across the component

Ohm's law
The special case of conduction through a metal is summed up in **Ohm's law**:

The current through a metallic conductor is proportional to the potential difference across the conductor provided the temperature remains constant.

Resistivity

The resistance of a component describes how well (or badly) a particular component or metal wire conducts electricity. It is often useful to describe the behaviour of a material; to do this we use the idea of **resistivity**.

The **resistance** of a wire is:
- directly proportional to its length, $R \propto L$
- inversely proportional to the cross-sectional area, $R \propto 1/A$

So, $R \propto \dfrac{L}{A}$

Hence:

$$R = \dfrac{\rho L}{A}$$

where ρ is the constant of proportionality, which is called the resistivity.

The units of resistivity are $\Omega\,\text{m}$, $[\rho = RA/L \rightarrow \Omega \times \text{m}^2/\text{m} = \Omega\,\text{m}]$

Worked example

A student wants to make a heating coil that will have a power output of 48 W when there is a potential difference of 12 V across it. The student has a reel of nichrome wire of diameter 0.24 mm. The resistivity of nichrome is $1.3 \times 10^{-8}\,\Omega\,\text{m}$.

Calculate the length of wire that the student requires.

Answer
The resistance of the coil can be found from the equation $P = V^2/R$.

$R = V^2/P = 12^2/48 = 144/48 = 3.0\,\Omega$

$R = \rho L/A$

$A = \pi(d/2)^2 = \pi \times (0.24 \times 10^{-3}/2)^2 = 4.52 \times 10^{-8}\,\text{m}^2$

$L = RA/\rho = 3.0 \times 4.52 \times 10^{-8}/1.3 \times 10^{-8} = 10.4\,\text{m}$

Potential difference and e.m.f.

These two terms have similar but distinct meanings. We have already met potential difference. Remember that it is defined as the work done, or energy transferred, when unit charge moves between two points. The term e.m.f. is used where a source of energy (such as a cell) gives energy to unit charge. However, it is a little more precise than this. If you feel a battery after it has delivered a current for some time it is warm. This means that, as well as the battery giving energy to the charge, the charge is doing some work in the battery. When e.m.f is defined, this work is included.

e.m.f. is defined as numerically equal to the energy converted from other forms into electrical potential energy per unit charge.

Potential difference is numerically equal to the energy converted from electrical potential energy to other forms per unit charge.

> **Tip**
>
> The term e.m.f. originally stood for electromotive force. This is rather confusing as it has nothing to do with force! Nowadays, e.m.f. stands as a term on its own and the use of the full electromotive force is best avoided.

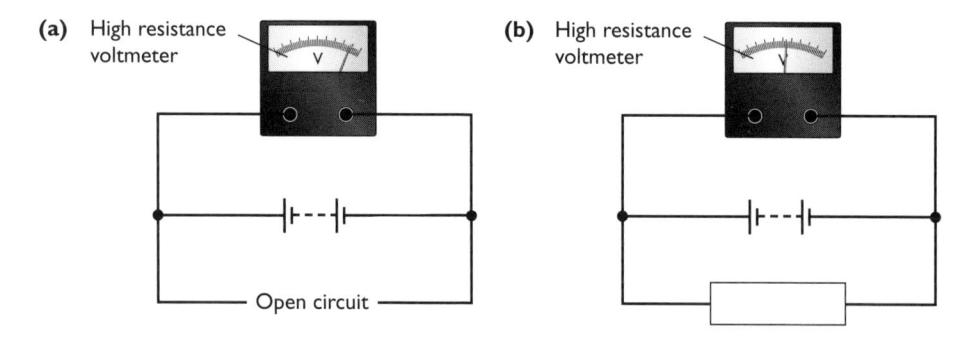

(a) High resistance voltmeter — Open circuit

(b) High resistance voltmeter

In the diagram above, circuit (a) shows the potential difference when (virtually) no current is taken from the battery. This is (almost) equal to the e.m.f.

Circuit (b) shows how the potential difference across the cell falls when a current is taken from it. Some work is done driving the current through the battery.

Internal resistance

We have seen how a source of e.m.f. has to do some work in driving a current through the source itself. In the case of a battery or cell, this is due to the resistance

V Electricity and magnetism

of the electrolytic solutions in the cell. In the case of a generator or transformer it is due to the resistance of the coils and other wiring in the apparatus. It is clear that the source itself has a resistance; this is called the **internal resistance** of the source.

It is often easiest to think of the two parts of a source of e.m.f. — the energy giver and the internal resistance — quite separately.

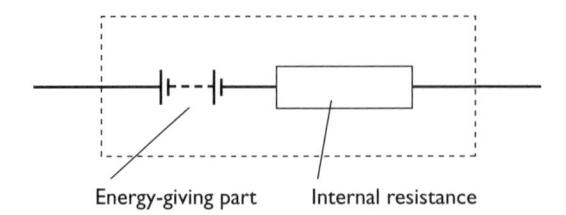

Energy-giving part   Internal resistance

Consider a battery of e.m.f. E and internal resistance r, driving a current through an external resistance R. The potential difference across the terminals of the battery is V.

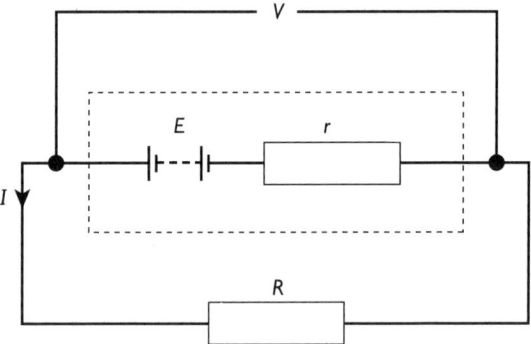

When you work with internal resistances treat them exactly the same as resistances in any other circuit. Work through the following equations to ensure that you understand the relationships.

$$E = I(R + r) = IR + Ir$$

but $IR = V$ and therefore

$$E = V + Ir$$

Worked example
A battery is connected across a resistor of 6.0 Ω and an ammeter of negligible resistance. The ammeter registers a current of 1.5 A. When the 6.0 Ω resistor is replaced by an 18 Ω resistor, the current falls to 0.6 A.

Calculate the e.m.f. and internal resistance of the battery.

Answer
Consider the 6.0 Ω resistor:
$E = IR + Ir = (1.5 \times 6.0) + 1.5r \rightarrow E = 9.0 + 1.5r$

Consider the 18 Ω resistor:
$E = IR + Ir = (0.6 \times 18) + 0.6r \rightarrow E = 10.8 + 0.6r$

Substitute for E in the second equation:
$9.0 + 1.5r = 10.8 + 0.6r$, therefore
$r = 2 \Omega$

Substitute for r in the first equation:
$E = 9.0 + (1.5 \times 2) = 12\,V$

Power output

The power output form a source is dependent on the internal resistance of the source and on the resistance in the external circuit (the load resistance). The graph shows how the power output varies. Note that there is maximum power output when the load resistance is equal to the internal resistance.

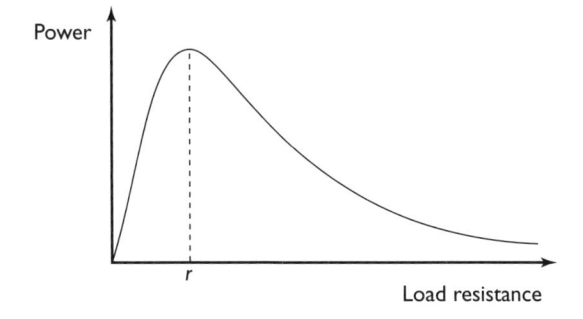

d.c. circuits

Signs and symbols

You should familiarise yourself with the circuit symbols.

Kirchhoff's laws

First law

The sum of the currents entering any junction in an electric circuit is equal to the sum of the currents leaving that junction.

This is a restatement of the law of conservation of charge. It means that the current going into a point is equal to the current leaving that point.

Worked example

Calculate the current I in the diagram.

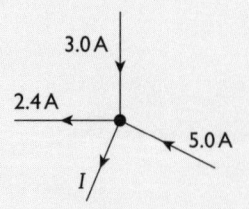

Answer

Consider the currents going into the point as positive and those leaving the point as negative.

$3.0 - 2.4 - I + 5.0 = 0$, therefore $I = 5.6$ A

Second law

In any closed loop in an electric circuit, the algebraic sum of the electromotive forces is equal to the algebraic sum of the potential differences

This is restatement of the law of conservation of energy. Remember that potential difference between two points is the work done per unit charge in moving from one point to the other. If the start point and the end point are the same then the net energy change, or work done, must be zero.

Going round a loop, we consider instances where energy is given to the charge to be positive and where energy is lost by the charge to be negative.

Worked example

The diagram shows a circuit. Calculate the e.m.f. of cell E_2 for the current through the ammeter to be zero.

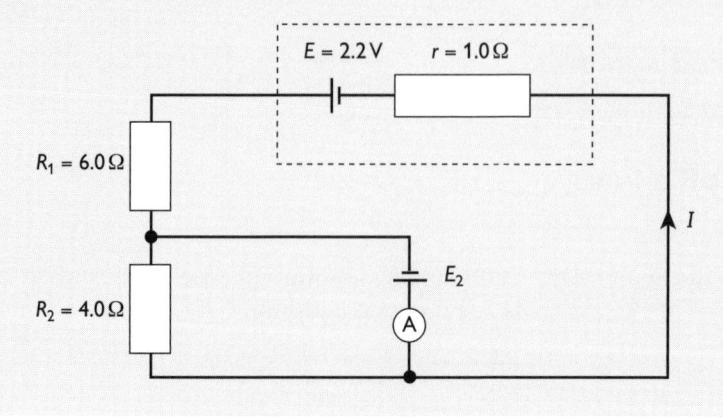

Answer
Consider the outer loop and move anticlockwise round the loop:
2.2 − 6.0I − 4.0I − 1.0I = 0 therefore:
I = 0.2 A

Consider the inner loop, which contains the 4.0 Ω resistor and the cell, E_2. Again move anticlockwise round the loop.

(−4.0 × 0.2) − E_2 = 0, therefore:
E_2 = −0.8 V

The minus sign shows that in order to satisfy the conditions the cell should be connected the other way round.

Hint

The e.m.f. of the second cell is put as −E_2 because the movement is from the positive to the negative cell — from a position of high potential energy to one of lower potential energy.

Resistors in series

To find the total resistance of resistors connected in series we can use Kirchhoff's second law.

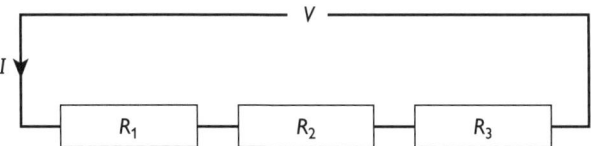

Going round the circuit:

$V - IR_1 - IR_2 - IR_3 = 0$

$V = IR_1 + IR_2 + IR_3 = I(R_1 + R_2 + R_3)$. But $V/I = R_{total}$, so

$R_{total} = R_1 + R_2 + R_3$

Resistors in parallel

To find the total resistance of resistors connected in parallel we can use Kirchhoff's laws.

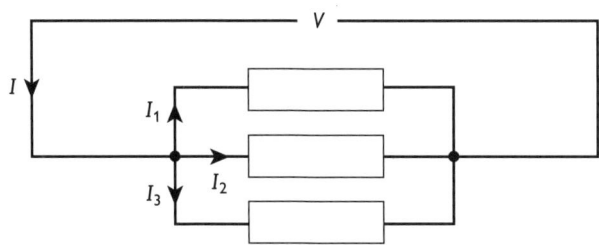

Using Kirchhoff's second law we can see there is the same potential difference across each of the resistors, therefore:

$I = V/R_{total}$, $I_1 = V/R_1$, $I_2 = V/R_2$ and $I_3 = V/R_3$

Using Kirchhoff's first law:

$I - I_1 - I_2 - I_3 = 0 \rightarrow I = I_1 + I_2 + I_3$, therefore:

$V/R = V/R_1 + V/R_2 + V/R_3$, and cancelling gives:

$1/R = 1/R_1 + 1/R_2 + 1/R_3$

Worked example
The diagram shows a network of resistors made up of five identical resistors each of resistance R.

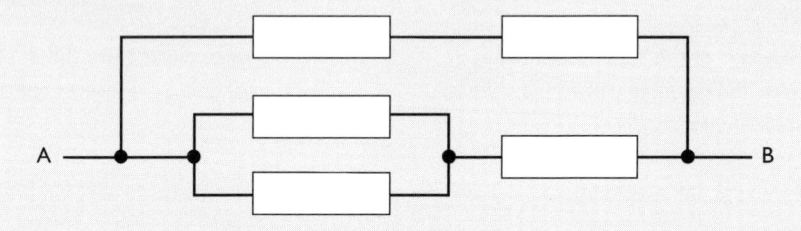

Calculate the resistance of the network.

Answer
resistance of the top line = $2R$
resistance of the pair of resistors in parallel = $(1/R + 1/R)^{-1} = 0.5R$
resistance of the lower line = $R + 0.5R = 1.5R$

$$\frac{1}{R_{total}} = \frac{1}{2R} + \frac{1}{1.5R} = \frac{3+4}{6R} = \frac{7}{6R}$$

therefore:
$R_{total} = 6R/7 = 0.86R$

Potential dividers

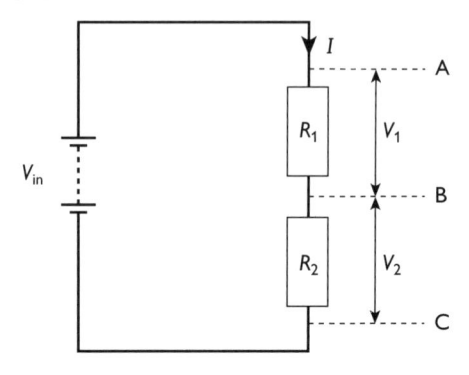

A potential divider does exactly what the name suggests. Study the diagram above. If there is a potential V across AC then the total potential drop is divided between AB and BC.

In the diagram, $V_1 = IR_1$ and $V_2 = IR_2$

Therefore:

$$\frac{V_1}{V_2} = \frac{IR_1}{IR_2} = \frac{R_1}{R_2}$$

A useful alternative way of working with this is:

$$V_{out} = \frac{R_2}{R_1 + R_2} \times V_{in}$$

where V_{out} is the potential drop across R_2 and V_{in} is the potential difference across the two resistors.

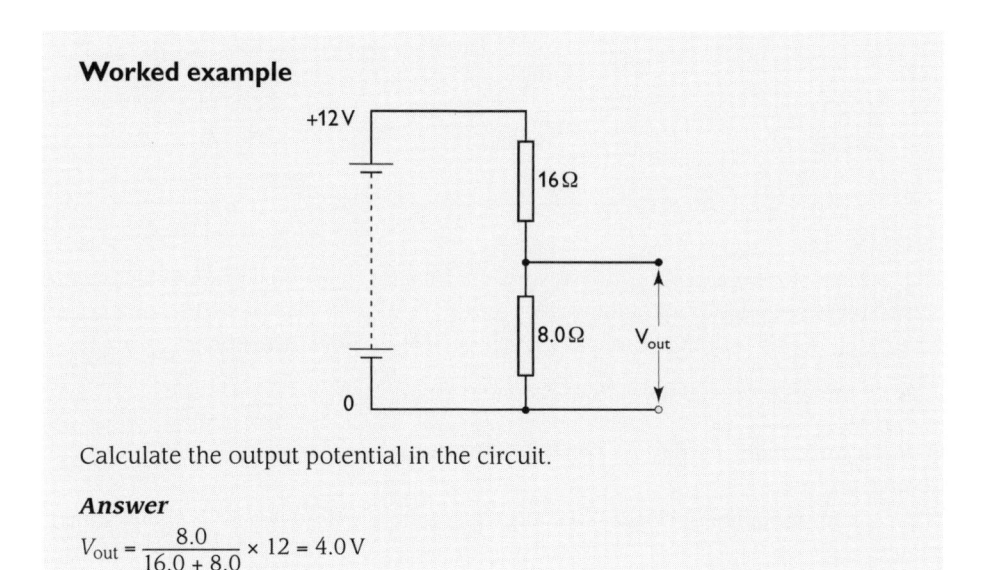

Worked example

Calculate the output potential in the circuit.

Answer

$$V_{out} = \frac{8.0}{16.0 + 8.0} \times 12 = 4.0\,V$$

Using potential dividers as input devices in control circuits

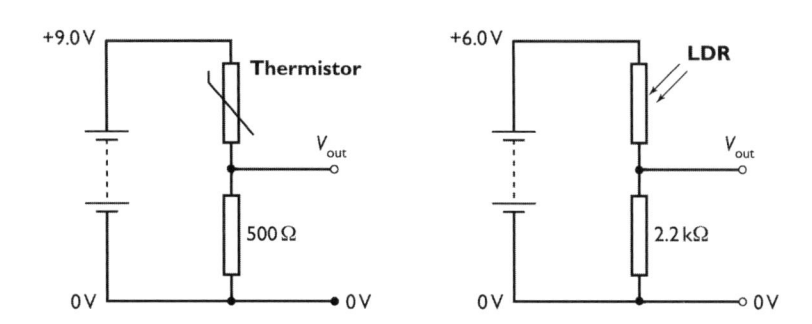

V Electricity and magnetism

Resistor R_1 is replaced by a thermistor. The thermistor resistance decreases when the temperature increases, so the output voltage V increases. This output voltage can then be, for example, the input to a circuit that switches on a fan, or one that can turn off a heater.

If R_1 is replaced by a light-dependent resistor (LDR), the resistance of the LDR increases when it gets dark. The output voltage would therefore decrease. If it is to be used to switch on a lamp this output must be first connected to an inverting circuit, and then to the lamp.

Using a potential divider to provide a variable voltage output

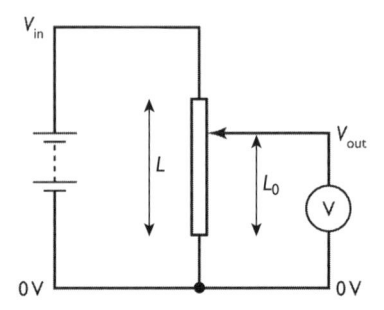

The two resistors are replaced by a single conductor, with a sliding contact to the conductor. The conductor could be a long straight wire, a strip of carbon or a coiled wire. Used in this way the potential divider is called a **potentiometer**.

If a uniform wire is used the output potential $V_{out} = \frac{L_0}{L} \times V_{in}$.

Using a potential divider to compare potential differences
Comparing cells
When a potential divider is used to compare potential differences it is usually called a potentiometer.

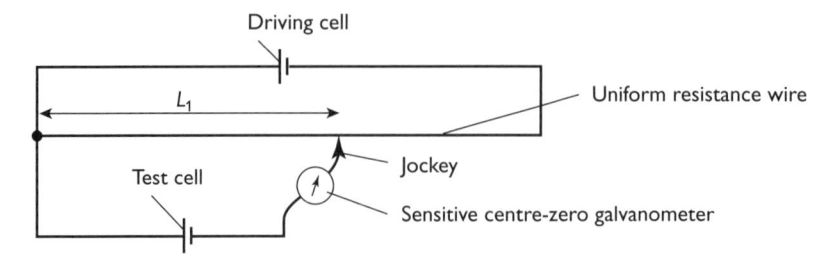

The circuit can be used to compare the e.m.f. of two cells. The position of the jockey is adjusted until the current through the galvanometer is zero. The e.m.f. (E_t) of the test cell is now equal to the potential drop across the length L_1 of the resistance wire. This method of measurement is known as a **null method**, null meaning nothing. The length L_1 is recorded.

The test cell is then replaced with a standard cell of e.m.f. E_s. The position of the jockey is adjusted until the new null reading is found. The new length (L_2) is measured and recorded.

The two e.m.f.s are related by the equation:

$$\frac{E_t}{E_s} = \frac{L_1}{L_2}$$

Comparing resistors

A similar method can be used to compare resistors. Two resistors are set up in series with a cell.

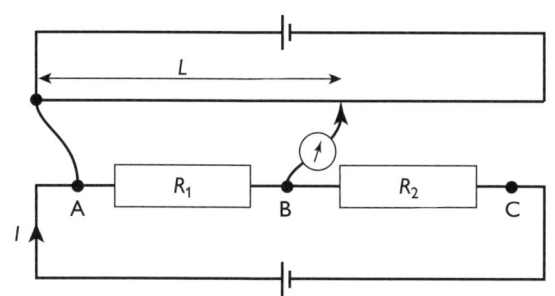

The series circuit is then connected to the potentiometer as in the diagram and the balance point is found ($L = L_1$). The potential drop across the resistor = IR_1.

The leads from the potentiometer are disconnected and then reconnected across the second resistor (points B and C on the diagram). The new balance point is found ($L = L_2$). The potential drop across this resistor = IR_2.

Hence:

$$\frac{R_1}{R_2} = \frac{L_1}{L_2}$$

> **Worked example**
>
> A potentiometer, which has a conducting wire of length 1.0 m, is set up to find the e.m.f. of a dry cell. When the dry cell is connected to the potentiometer, the balance length is found to be 43.5 cm. A standard cell of e.m.f. 1.02 V is used to replace the dry cell. The balance length is now 12.9 cm less than for the dry cell. Find the e.m.f. of the dry cell.
>
> **Answer**
> The balance length for the standard cell = 43.5 − 12.9 = 30.6 cm
>
> $$\frac{E_t}{E_s} = \frac{L_1}{L_2}$$
>
> $$\frac{E_t}{1.02} = \frac{43.5}{30.6} = 1.42$$
>
> $E_s = 1.42 \times 1.02 = 1.45$ V

Modern physics
Nuclear physics
Nuclear model of the atom
Rutherford scattering experiment

By the early part of the twentieth century, following the discovery of the electron in 1896, it was recognised that the atom had structure. Early models of the atom considered it to be a positive cloud of matter with electrons embedded in it — the plum-pudding model.

In 1911, Rutherford's alpha scattering experiment led to a model of the atom with a positively charged **nucleus** containing all the positive charge and virtually all the mass of the atom. This nucleus is surrounded by the much smaller, negatively charged **electrons**.

A plan view of the Rutherford alpha scattering apparatus is shown below.

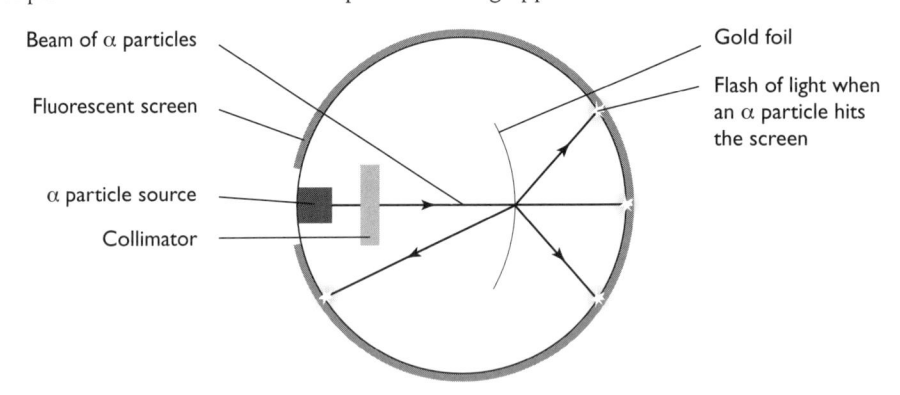

The following diagram shows deflection of alpha particles by the nucleus.

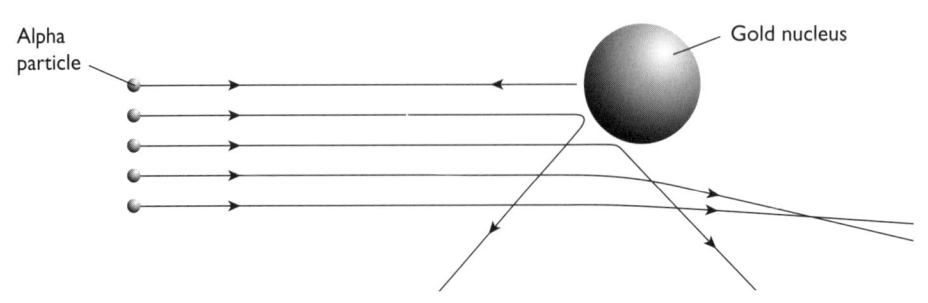

Results from the alpha scattering experiment
- The vast majority of the particles passed straight through the foil with virtually no deflection.
- A few (approximately 1 in 10 000) of the alpha particles were deflected through angles in excess of 90°.

Alpha particles are positively charged with a mass about 8000 times that of an electron.

The large-angle deflection of the alpha particles could only occur if the alpha particles interacted with bodies more massive than themselves. This led Rutherford to develop the solar-system model of the atom:

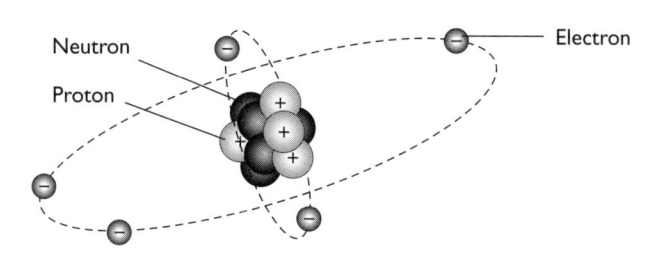

Rutherford's solar-system model of the atom

The small numbers that are deflected through large angles indicate that the nucleus is very small. The proportions deflected in different directions enabled Rutherford to estimate the diameter of the nucleus as being in the order of 10^{-14} m to 10^{-15} m. This compares with an atomic diameter of about 10^{-10} m.

The diagram is not drawn to scale. If it were, and the nucleus was kept to this size, then the electrons would be over 100 m away!

Subatomic particles

Subatomic particles are shown in Table 12. You can see that the nucleus has a structure, being made up of protons and neutrons.

Table 12

Particle	Charge*	Mass**	Where found
Proton	+1	1	In the nucleus
Neutron	0	1	In the nucleus
Electron	−1	1/1840	In the outer atom

*Charge is measured in terms of the electronic charge (e): e = 1.6×10^{-19} C.
**Mass is measured in unified atomic mass units (u): 1 u is 1/12 of the mass of a carbon-12 atom.

The chemical elements

The different elements and their different chemical properties are determined by the number of protons in the nucleus which, in turn, determines the number of electrons in the outer atom. The different atoms, or more precisely their nuclei, are fully described by the number of protons (the **proton number**) and the total number of neutrons plus protons (the **nucleon number**) in the nucleus.

> **Tip**
>
> You might see the proton number referred to as the atomic number and the nucleon number as the mass number.

All elements have **isotopes**. Isotopes have the same number of protons (and hence the same chemical properties) but different numbers of neutrons.

A **nuclide** is a one type of nucleus with a specific nucleon number and a specific proton number. The notation we use to describe a nuclide is:

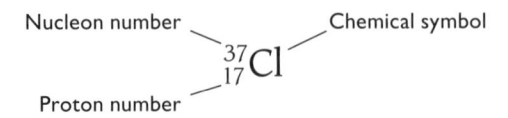

Nucleon number — $^{37}_{17}\text{Cl}$ — Chemical symbol

Proton number

This nuclide is an isotope of chlorine. The nucleus contains 17 protons and (37 − 17) = 20 neutrons. This isotope makes up about 24% of naturally occurring chlorine. The other 76% is made up of the isotope $^{35}_{17}\text{Cl}$, which contains 18 neutrons.

Radioactive decay

Some nuclides are unstable and decay by emitting radiation; this is known as radioactive decay. The rate of radioactive decay is not dependent on outside conditions (e.g. temperature, pressure). In this sense, the decay is said to be **spontaneous**. It is only dependent on the stability of the particular nuclide. However, if only a single nucleus is considered it is impossible to predict *when* it will decay. In this way, radioactive decay is **random**. However, we can say that there is a fixed chance of decay occurring within time Δt. Hence, a fixed proportion of a sample containing millions of atoms will decay in that time interval. This randomness is clearly demonstrated by the fluctuations in the count rate observed when radiation from a radioactive isotope is measured with a Geiger counter or other detector.

The three common types of radioactive decay — alpha decay, beta decay and gamma decay — are shown in Table 13.

Table 13

Name of particle	Nature	Charge	Penetration	Relative ionising power	Reason for decay
Alpha (α)	Fast moving helium nucleus	+4e	Very low; stopped by thin card or aluminium foil	High (10^4)	Nucleus too large; helium groupings form within the nucleus and sometimes escape
Beta (β)	Very fast moving electron	−e	Fair; stopped by several mm of aluminium	Fair (10^2)	Nucleus has too many neutrons; a neutron decays into a proton and an electron; the electron escapes from the nucleus
Gamma (γ)	Short wavelength electromagnetic radiation	Zero	High; only partially reduced by several cm of lead	Low (1)	Usually emitted in conjunction with an alpha decay as the new nucleus drops into a more stable, lower energy state

International AS and A Level Physics Revision Guide

The different penetrating powers of the radiations can be explained by the relative ionising powers. Each time radiation ionises a particle it loses energy. Thus alpha particles, which cause many ionisations per unit length, lose their energy in a much shorter distance than gamma rays, which cause far fewer ionisations.

Deflection of the radiations in electric fields

Both alpha and beta particles are charged and are consequently deflected by both electric and magnetic fields. Gamma rays are uncharged and are, therefore, not deflected in either type of field. Although alpha particles have twice the charge of beta particles, they are much more massive and are therefore deflected by much less. Because the charge on an alpha particle is positive and that on a beta particle is negative, they are deflected in opposite directions. Alpha particles from a particular source all have the same energy. Consequently, in the same field they all deflect by the same amount. Beta particles have a range of energies, so they have a range of different deflections.

Equations of radioactive decay

Alpha decay

$$^{241}_{95}Am \rightarrow ^{237}_{93}Np + ^{4}_{2}\alpha + \text{energy}$$

Here, americium decays into neptunium. Note how the total of the two proton numbers is the same after the decay as before it. Likewise, the total nucleon numbers are unaltered in the decay. This is sometimes expressed as 'conservation of the proton number' and 'conservation of the nucleon number'. Conservation of the proton and nucleon numbers is true for all nuclear reactions.

Beta decay

$$^{90}_{38}Sr \rightarrow ^{90}_{39}Y + ^{0}_{-1}\beta + \text{energy}$$

The beta particle is considered to have a proton number of –1, because it has a charge of –1 proton charge. Note again the conservation of proton and nucleon numbers.

Other forms of decay

Although alpha, beta and gamma decay are the most common forms of decay, there are many other possibilities. An important example is when a neutron is absorbed by a nitrogen nucleus, which then decays by emitting a proton.

$$^{14}_{7}N + ^{1}_{0}n \rightarrow ^{14}_{6}C + ^{1}_{1}p$$

> **Worked example**
>
> A $^{16}_{8}O$ nucleus absorbs a neutron. The newly formed nucleus subsequently decays to form a $^{17}_{9}F$ nucleus.
>
> (a) Write an equation to show the change when the neutron is absorbed.
>
> (b) Deduce what type of particle is emitted when the decay of the newly formed nucleus occurs.

Answer

(a) $^{16}_{8}O + ^{1}_{0}n \rightarrow ^{17}_{8}O$

(b) new proton number = 9, old proton number = 8

new nucleon number = 17, old nucleon number = 17

Therefore a beta-particle ($^{0}_{-1}\beta$) is emitted.

Mass–energy

In his theory of relativity, Einstein introduced the idea that mass and energy are closely related. One way of viewing this is to consider energy as having mass. It is an established experimental fact that the mass of a particle travelling at near the speed of light is greater than its mass at rest. This idea can be extended to other forms of energy.

When a radioactive nucleus decays it drops into a lower energy state. The average energy per nucleon is slightly less in the new nucleus than in the original nucleus. Consequently, the average mass per nucleon is less. The energy that is 'lost' is taken away by the alpha or beta particle (as kinetic energy) or the gamma ray (electromagnetic energy). These have more energy and hence more mass. Consequently, the total mass and the total energy are conserved. Put another way, the mass–energy is conserved.

This is explored further in the A2 course.

AS Experimental skills and investigations

Almost one-quarter of the marks for the AS examination are for experimental skills and investigations. These are assessed on Paper 3, which is a practical examination.

There is a total of 40 marks available on this paper. Although the questions are different on each Paper 3, the number of marks assigned to each skill is always the same. This is shown in the table below.

Skill	Total marks	Breakdown of marks	
Manipulation, measurement and observation, MMO	16 marks	Successful collection of data	13 marks
		Range and distribution of values	1 mark
		Quality of data	2 marks

International AS and A Level Physics Revision Guide

Presentation of data and observations, PDO	10 marks	Table of results: layout	1 mark
		Table of results: raw data	1 mark
		Table of results: calculated quantities	2 marks
		Graph: layout	1 mark
		Graph: plotting of points	1 mark
		Graph: trend line	1 mark
		Display of calculation and reasoning	3 marks
Analysis, conclusions and evaluation, ACE	14 marks	Interpretation of graph	2 marks
		Drawing conclusions	3 marks
		Estimating uncertainties	1 mark
		Identifying limitations	4 marks
		Suggesting improvements	4 marks

The syllabus explains each of these skills in detail, and it is important that you read the appropriate pages in the syllabus so that you know what each skill is and what you will be tested on.

There is a great deal of information for you to take in and skills for you to develop. The only way to do this really successfully is to do lots of practical work and gradually build up your skills bit by bit. Don't worry if you don't get everything right first time. Just take note of what you can improve next time — you will steadily get better and better.

The examination questions

There are usually two questions on Paper 3. The examiners will take care to set questions that are not exactly the same as any you have done before.

Read the question carefully before you start. It is important that you follow the instructions and do exactly what the question asks. It is easy to jump in, thinking that you have done this type of practical before, when what you are being asked to do is subtly different.

Question 1

This question is a complete experiment. For example, it might be:
- investigating factors affecting the period of an oscillating system
- investigating the effect of forces acting on a system
- investigating an electric circuit

This experiment will ask you to:
- set up some apparatus or connect up a circuit
- take a series of readings with the apparatus
- use the raw data to find derived data
- plot a graph using the raw or derived data

- find information from the graph, such as its gradient and/or the *y*-intercept
- use the information from the graph to calculate the constants in an equation

Question 2

This question takes the starting point of an investigation. You will be given some simple apparatus, which you may have to assemble. You will then:
- take a couple of initial measurements in which you will have to justify the number of significant figures used, or explain how you set the apparatus up to ensure as accurate measurement as possible
- take measurements of the independent and dependent variables and estimate the absolute and/or percentage uncertainty in one of the readings
- take a second pair of readings and justify whether or not the readings you have taken support a particular hypothesis
- identify limitations or sources of error in the experiment
- suggest improvements that would make the experiment more reliable

> **Tips** Before the exam:
> - Every time you do a practical during your AS course, time yourself. Are you working quickly enough? You will probably find that you are slow to begin with but as the course progresses try to work a little faster as your confidence improves.
>
> In the exam:
> - Read the question carefully and do exactly what it asks you to do. This is unlikely to be exactly the same as anything you have done before.
> - Leave yourself enough time to complete both questions, spending approximately 60 minutes on each.

How to get high marks in Paper 3

It is impossible to predict what sort of experiment you will have to do in the examination. However, many of the experiments are variations on experiments that you will have met during your course. You need to make the most of your practical sessions in order to develop the skills that will make the practical examination easier.

Making measurements

Physics is a science of measurement. You need to be familiar with the use of a range of instruments: millimetre scales, micrometers, vernier scales, measuring cylinders, balances, newton meters, thermometers, ammeters, voltmeters, cathode-ray oscilloscopes. You will have become familiar with many of these instruments in your previous courses. Nevertheless, it is important to ensure that you are competent in the use of all of these. If you are not sure, refer to the relevant part of this book or ask your teacher for help.

You need to be able to assemble basic apparatus according to the instructions given. There will be nothing too difficult but the more practice you have had the easier you will find this, and the quicker you will be able to do it. One of the skills you need to develop is the ability to build electric circuits. Difficulties often arise when there are parallel circuits. The simple rule here is to build the main series part of the circuit and then add the parallel arms afterwards.

The stages in building a circuit are shown in the diagram on page 104.

Stage 1 Connect the positive terminal of the cell to the red terminal of the ammeter.

Stage 2 Connect the black terminal of the ammeter to the first resistor.

Stage 3 Complete the main circuit by connecting the second resistor to the first, and then to the negative terminal of the cell.

Stage 4 Connect two leads to the voltmeter and then connect them to the relevant points in the circuit. Note the red terminal of the voltmeter is nearest to the positive terminal of the cell.

Tip

If you have difficulty in setting up the apparatus, or connecting a circuit, inform the supervisor. There might be a problem with your apparatus, in which case there will be no penalty (you might even get a little extra time). If there is nothing wrong with the apparatus, the supervisor will make a note on your paper that help was required in setting it up and you might lose 1 or 2 marks. However, better to lose the odd mark here than to lose everything because you were unable to proceed.

Variables

The factor that you change or control is called the **independent variable**. The factor that is affected and that you measure when you collect your results is the **dependent variable**. The table shows some examples.

	Investigation	Independent variable	Dependent variable
1	Investigating the height of a bouncing ball	Height from which the ball is dropped	Height to which the ball bounces
2	Investigating the period of vibration of masses suspended by a spring	Mass on the end of the spring	Periodic time
3	Investigating the melting of ice in water	Temperature of water	Time taken to melt
4	Investigating the current through resistors	Resistance of resistor	Current
5	Investigating e.m.f. using a potentiometer	e.m.f.	Balance length

We will refer to these examples later in the text, so you might like to put a marker on this page so that you can easily flip back as you read.

AS Experimental Skills and Investigations

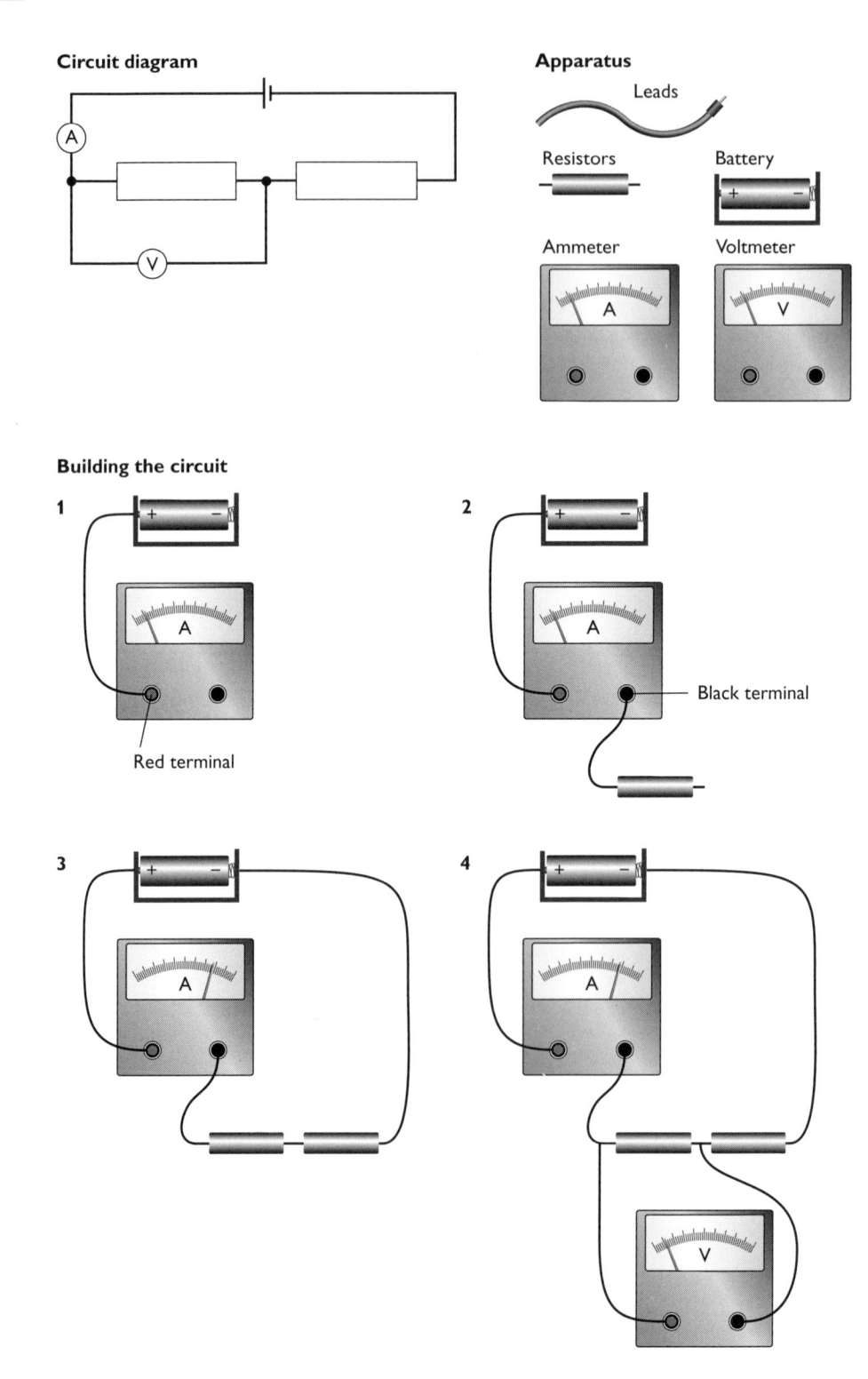

Circuit diagram

Apparatus

Leads

Resistors

Battery

Ammeter

Voltmeter

Building the circuit

1

Red terminal

2

Black terminal

3

4

At A2, you will need to consider variables other than those you have chosen that might affect your results. These are known as **control variables** and we will consider them in detail in the A2 section of the book.

Range of readings

When you plan your experiment you should use as wide a range of values for the independent variable as possible. If you consider Investigation 3 — the melting ice experiment — the range of temperatures of the water in the beaker should be from nearly 100 °C to about 10 °C. You will probably be told how many readings to take but it likely to be a minimum of six sets. The values chosen for the independent variable should be taken at roughly equal intervals. A sensible spread might be 95 °C, 80 °C, 60 °C, 45 °C, 30 °C and 15 °C.

It sometimes makes sense to take several readings near a particular value — for example, if the peak value of a curved graph is being investigated. Practice in carrying out experiments will give you experience in deciding if this type of approach is necessary.

Accuracy, precision and uncertainty

Precision and accuracy are terms which often cause confusion. **Accuracy** is how close to the 'real value' a measurement is. Consider a rod of diameter 52.8012 mm. Suppose that you use a ruler and measure it to be 53 mm. This is accurate but it is not very precise. If your friend uses a micrometer screw gauge and measures it as 52.81 mm this is much more precise, even though the final figure is not totally accurate.

No measurement can be made to absolute precision — there is always some **uncertainty**. We can describe the uncertainty as the range of values in which a measurement could fall. If a result is recorded as 84.5 s, this implies that there is an uncertainty of at least 0.1 s, perhaps more. You may see such a reading written as 84.5 ± 0.2 s. The 0.2 s in this reading is called the **absolute uncertainty**.

It is often convenient to express an uncertainty as a percentage of the reading. This is known as the **percentage uncertainty**.

   percentage uncertainty = (absolute uncertainty/reading) × 100%

The percentage uncertainty in the previous example is:

   (0.2/84.5) × 100% = 0.24%

Precision of measurement

When making a static measurement (for example, the length of a pendulum) you should normally measure to the nearest division on the instrument. The exception to this is if the divisions are more than one millimetre apart. In this case, you need to judge to the nearest half division or better. When making a dynamic measure (for example, the height to which a ball bounces), then other considerations come into play — the ball is moving, so you have to judge when it is at its maximum height. This is a much more difficult task. You can probably measure only to the nearest 5 millimetres or worse.

Many digital stopwatches measure to 1/100 of a second. However, the uncertainties in the reaction times of manually starting and stopping a stopwatch are much greater than this. The best you can manage is to measure to the nearest 1/10 of a second. Until 1977, world records for running events were given to only this precision. It was only with the advent of electronic timing that it became possible to record them to 1/100 of a second. The current world record for the men's 100 m is 9.58 s. This suggests an absolute uncertainty of ±0.01 s, a percentage uncertainty of approximately 0.1%. This has the knock-on effect that for the world record to be valid the track must also be measured to a precision of 0.1% or better. This means an absolute uncertainty of 10 cm.

Repeating readings

You should recognise that you can reduce the chances of serious error by repeating readings, all of which should be included in your records. In general, it is only necessary to repeat those readings with the potential for the highest percentage uncertainty.

> **Worked example**
>
> A student is measuring the relationship between the height from which a ball is dropped and the height to which it bounces. The student drops the ball from a height of 1.0 m and it bounces to a height of approximately 50 cm.
>
> State which measurements should be repeated and justify your answer.
>
> *Answer*
>
> The height from which the ball is dropped can be measured to the nearest millimetre, giving a percentage uncertainty of 0.1%. The height to which the ball bounces can only be estimated to about 1 cm, giving a percentage uncertainty of 2%. This is a much larger percentage uncertainty. Hence the height of bounce should be repeated but the height from which the ball is dropped need not be repeated.

Types of error

We can loosely put errors in measurement into one of two categories — **random** errors and **systematic** errors.

Random errors occur due to a lack of precision in taking readings, slight changes in experimental conditions and making value judgements when taking measurements. Where it is felt that the random error might be significant, repeated readings should be taken. These readings will give you further information about the uncertainty in the measurement. For instance, if you take five readings of the maximum amplitude of a pendulum as 24.1 cm, 23.8 cm, 24.3 cm, 23.6 cm and 24.0 cm, this gives an average value of 23.96 cm, which would be rounded to 24.0 cm. It is quite clear that the measuring instrument can measure to the nearest millimetre, but to give the reading as 24.0 ± 0.1 cm would be claiming a greater precision than you have. The largest deviation from the average value is 0.4 cm, so the correct precision is ± 0.4 cm. The reading should be recorded as 24.0 ± 0.4 cm.

Systematic errors generally occur because of faults in a measuring instrument, or are repeated errors such as a parallax error (not looking perpendicularly at a measuring instrument but always at the same angle). It is worth noting that if the angle from which the instrument is viewed changes, then the error introduced will be random, not systematic. Careful thought when setting up and carrying out the experiment should ensure that parallax errors are avoided.

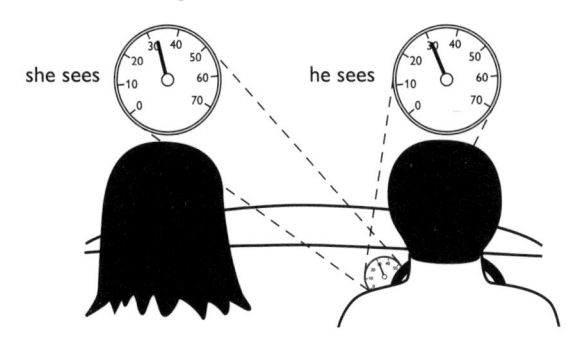

In the diagram above, the passenger thinks that the driver is travelling faster than he really is because she is looking at the speedometer at an oblique angle, introducing a parallax error.

The most common form of systematic error due to a faulty instrument is a zero error. When you take a micrometer screw gauge and close the jaws using the ratchet, you should check if the zero is lined up correctly. If it is not, then this is easy to rectify by simply subtracting the error from the reading. (Don't forget that if the error is a minus quantity, subtracting a minus quantity means adding its magnitude to the measured quantity.)

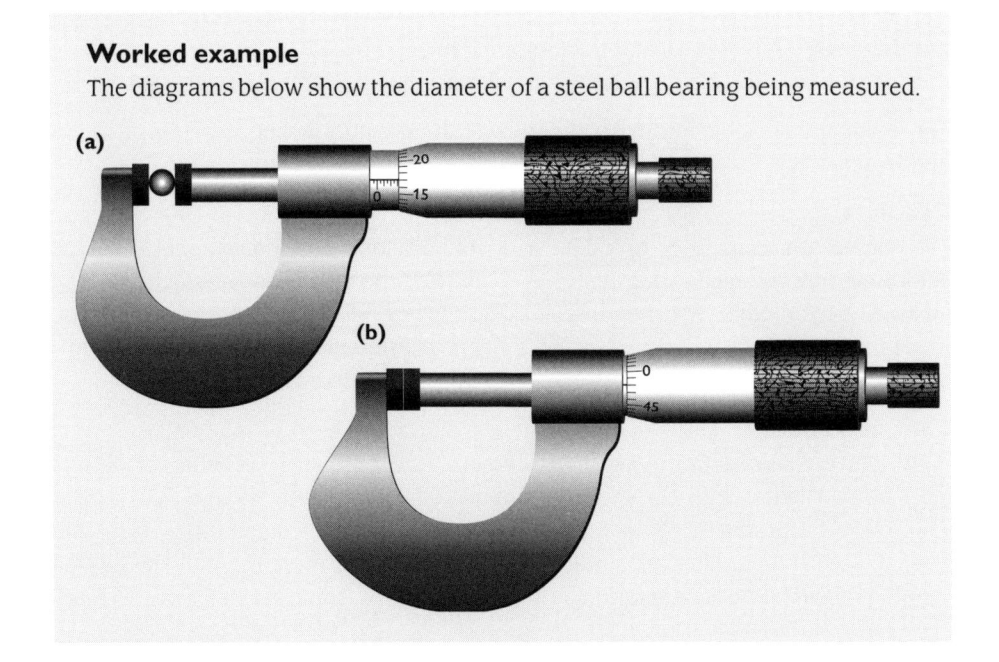

> **Worked example**
> The diagrams below show the diameter of a steel ball bearing being measured.
>
> (a)
>
> (b)

> Determine the diameter of the ball bearing.
>
> *Answer*
> reading on the barrel = 3.00 mm
> reading on the thimble = 0.17 mm
>
> calculate the final reading by subtracting the zero error from the intial reading:
> reading on micrometer = 3.17 mm
> zero error = −0.02 mm
> diameter of the ball bearing = 3.17 − (−0.02) = 3.19 mm

The other type of systematic error you could encounter is an instrument with a wrongly calibrated scale — that is, it consistently reads high or low, at a steady percentage of the true reading. An example might be a stopwatch that runs slow. This is more difficult to allow for unless you have a standard with which to compare it. It will not cause any scattering of the points on a graph but it will cause a shift in the gradient of the graph.

Presentation of data and observations

You should be in the habit of recording your measurements and results directly into your laboratory notebook, rather than using scraps of paper which might get lost or destroyed. This means you need to be organised. You must think clearly, before you start your work:
- What measurements do I need to make?
- What measurements do I need to repeat?
- What quantities do I need to calculate from my raw results?

You then need to draw a table that has sufficient columns and rows to accommodate these quantities, including columns for repeated readings and their averages.

The heading for each column should include the quantity being measured and the unit in which it is measured.

Raw data
The degree of precision of raw data in a column should be consistent. It will be determined by the measuring instrument used or the precision to which you can measure. This means that the number of significant figures may *not* be consistent. An example might be when measuring across the different resistors using a potentiometer, where the balance points may vary from 9.3 cm to 54.5 cm.

Calculated data
With data calculated from raw measurements the number of significant figures must be consistent with the raw measurements. This usually means that, except where they are produced by addition or subtraction, calculated quantities should be given to the same number of significant figures as (or one more than) the measured quantity of least precision. If a time is measured as 4.1 s, squaring this gives 16.81 s^2.

However, you would record the value as either 16.8 s² (or perhaps 17 s²). Again the number of significant figures in the column is not necessarily consistent.

The table below shows some readings from a potentiometer experiment and demonstrates how readings should be set out:
- the column headings, with quantity and unit
- the raw data to the same precision
- the calculated data to the relevant number of significant figures

R/Ω	L_1/m	L_2/m	L_{av}/m	$\ln(L_{av})$
47	0.191	0.194	0.193	−1.65**
100	0.381	0.379	0.380*	−0.968
220	0.778	0.784	0.781	−0.247

* Do not forget to include the zero, to show that the length has been measured to the nearest millimetre.
** You could justify writing this as −1.645, one more significant figure than the raw data. If this helped to plot a more precise graph, then it would be sensible to do this.

Graphs
Reasons for plotting graphs
Graphs:
- tend to average data thereby reducing the effects of random errors
- identify anomalous points (which should then be investigated further)
- tend to reduce the effect of systematic errors
- give information that can be used to identify relationships between variables

Rules for plotting graphs
1 **Draw and label axes**. Axes should be labelled with the quantity and the unit in a similar manner to column headings in a table. In general, the independent variable (the one you control) is put on the horizontal axis (x-axis). The dependent variable (the one that changes due to changes in the independent variable) goes on the vertical axis (y-axis).
2 **Choose sensible scales**. Scales should be chosen so that the points occupy at least half the sheet of graph paper used. However, awkward scales (1:3, 1:7, 1:11, 1:13 or their multiples) must be avoided. You do not necessarily have to include the origin on the graph if this means that a better use of the graph paper can be achieved.
3 **Plot points accurately**. Points should be plotted by drawing a small cross with a sharp pencil. Do *not* use dots or blobs.
4 **Draw the best-fit straight line or best smooth curve**. When you draw a straight line use a 30 cm ruler and a sharp pencil. There should be an equal number of points above and below the line. Take care that those points above and those below the line are evenly distributed along the line.
5 **Identify and check any anomalous points**. If a point is well off the line, go back and check it. In all probability you will have made an error, either in plotting the point or in taking the reading.

AS Experimental Skills and Investigations

Do *not* fiddle your straight line or curve so that it goes through the origin. There may be good reasons why the dependent variable is not zero when the independent variable is zero. Consider Investigation 2 in the table on page 103 — the experiment to investigate the period of vibration of a mass on the end of a spring — the measurements of the mass on the spring do *not* make allowance for the mass of the spring itself.

A typical straight-line graph is shown below.

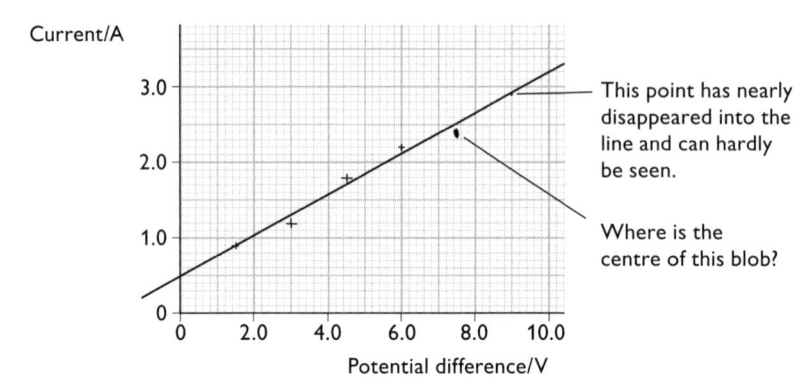

This point has nearly disappeared into the line and can hardly be seen.

Where is the centre of this blob?

Curves should be drawn with a single sweep, with no feathering or sudden jerks. You need to practise doing this.

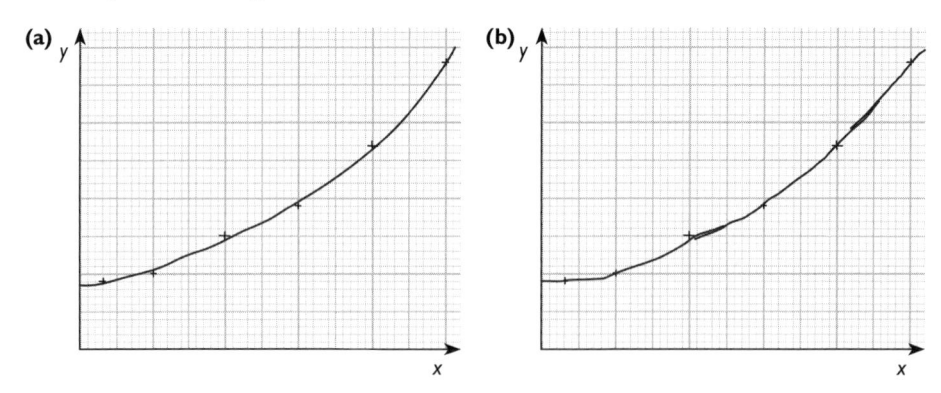

Graph (a) shows a well-drawn smooth curve. Graph (b) shows a poorly drawn curve through the same points. Note the jerkiness between the first two points and the feathering between points 2 and 4 and between points 5 and 6.

Measuring the gradient of a graph

The gradient of a graph is defined as

$$\frac{\text{Change in } y}{\text{Change in } x} = \frac{\Delta y}{\Delta x} = \frac{y_2 - y_1}{x_2 - x_1}$$

When choosing the points to calculate the gradient you should choose two points on the line (not from your table of results). To improve precision the two points should be as far apart as possible.

In the straight-line graph on page 110, two suitable points might be (0, 0.5) and (10.0, 3.2).

This gives a gradient $= \dfrac{3.2 - 0.5}{10.0 - 0} = 0.27 = $ A V^{-1}

Note that you should always include the unit of the gradient.

You may be asked to find the gradient of a curve at a particular point. In this case, you must draw a tangent to the curve at this point and then calculate the gradient of this line in a similar way to that described above.

Finding the y-intercept

The y-intercept of a graph is the point at which the line cuts the y-axis (that is when $x = 0$). In the example on page 110, the intercept is 0.5 A. When a false origin is used, it is a common mistake for students to assume that the vertical line drawn is the zero of x, so if you have used a false origin check carefully that the vertical line is at $x = 0$.

If, however, the chosen scale means that the y-intercept is not on the graph, it can be found by simple calculation.

- Calculate the gradient of the graph.
- The equation for a straight line graph is $y = mx + c$. Choose one of the points used for calculating the gradient and substitute your readings into the equation.

Worked example

The voltage input to an electrical device and the current through it were measured. The graph below was drawn from the results.

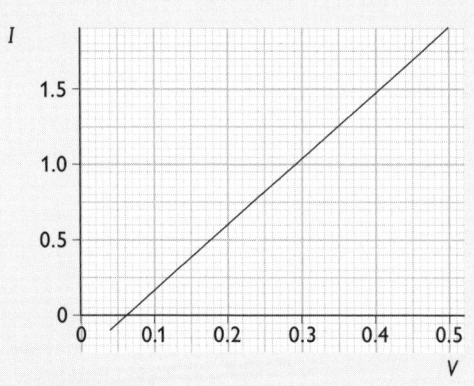

Determine the y-intercept on the graph.

Answer

Find the gradient. Use the points (0.06, 0) and (0.50, 1.90).

gradient $= \dfrac{\Delta y}{\Delta x} = \dfrac{y_2 - y_1}{x_2 - x_1} = \dfrac{1.90 - 0}{0.50 - 0.06} = 4.3$ AV^{-1}

Substitute the first point and the gradient into the equation $y = mx + c$.

$0 = (4.3 \times 0.06) + c$

$c = -0.26$ A

Evaluation of evidence

During an experiment you should record any uncertainties in your measurements.

It is important to note that questions on the combinations of uncertainties are often set in the theory papers. To find the uncertainty of a combination of variables, the rules are:
- for quantities that are added or subtracted, the absolute uncertainties are added
- for quantities that are multiplied together or divided, the *fractional* (or *percentage*) uncertainties are added
- for a quantity that is raised to a power, to calculate a final uncertainty the fractional uncertainty is multiplied by the power and the result is treated as a *positive* uncertainty

Worked example 1

The currents coming into a junction are I_1 and I_2. The current coming out of the junction is I. In an experiment the values of I_1 and I_2 are measured as 2.0 ± 0.1 A and 1.5 ± 0.2 A respectively.

Write down the value of I with its uncertainty.

Answer

$I = I_1 + I_2 = (2.0 \pm 0.1) + (1.5 \pm 0.2)$

The quantities are being added so to find the uncertainty the uncertainties of the original quantities are added.

Hence $I = 3.5 \pm 0.3$ A

Worked example 2

The acceleration of free fall g is determined by measuring the period of oscillation T of a simple pendulum of length L. The relationship between g, T and L is given by the formula $g = 4\pi^2 (L/T^2)$.

In the experiment, L was measured as 0.55 ± 0.02 m, and T as 1.50 ± 0.02 s.

Find the value of g and its uncertainty.

Answer

$g = 4\pi^2 (L/T^2) = 4\pi^2 (0.55/1.50^2) = 9.7 \text{ m s}^{-2}$

To find the uncertainties, the second and third rules are applied.

Fractional uncertainty in $L = 0.02/0.55 = 0.036$

Fractional uncertainty in $T = 0.02/1.50 = 0.013$

Fractional uncertainty in $T^{-2} = 2 \times 0.013 = 0.026$

International AS and A Level Physics Revision Guide

> Fractional uncertainty in g = fractional uncertainty in L + fractional uncertainty in T^{-2} = 0.036 + 0.026 = 0.062
>
> The absolute uncertainty in g = 9.7 × 0.062 = 0.6
>
> Thus $g = 9.7 \pm 0.6 \, \text{m s}^{-2}$

It is worth noting that the examiners are looking for the absolute uncertainty, not the percentage uncertainty. If you take the short cut and leave your answer as 9.7 ± 6.2%, you will lose credit. It is also worth noting that it is poor experimental practice to take only one reading and to try to find a value of g from that. You should take a series of readings of T for different lengths L, and then plot a graph of T^2 against L. The gradient of this graph would be equal to $4\pi^2/g$.

To ascertain if an experiment supports or fails to support a hypothesis, your result should lie within the limits of the percentage uncertainties. To support the hypothesis in the absence of any uncertainty calculations, a good rule of thumb is that the calculated value should lie within 10% of any predicted value.

The following worked examples take you through some of the stages of evaluating evidence.

> **Worked example 1**
> In an initial investigation into the time it takes for an ice cube to melt (Investigation 3 in the table on page 103) in a beaker of water the following results are obtained.
>
> Trial 1:
> initial temperature of the water = 50 °C
> time taken (t) to melt = 85 s
>
> Trial 2:
> initial temperature of the water = 80 °C
> time taken to melt = 31 s
>
> (a) Explain why it is only justifiable to measure the time taken for the ice cube to melt to the nearest second.
>
> (b) Estimate the percentage uncertainty in this measurement in trial 1.
>
> (c) Estimate the percentage uncertainty in this measurement in trial 2.
>
> (d) Why is it more important to calculate the uncertainty in the time rather than in the initial temperature of the water?

Answer

(a) Even though the stopwatch that was used may have measured to the nearest one-hundredth of a second, it was difficult to judge when the last bit of ice disappeared.

(b) Suppose that the absolute uncertainty = ±5 s

percentage uncertainty = $\pm \frac{5}{85} \times 100\% = 6\%$

(c) absolute uncertainty = ±5 s

percentage uncertainty = $\pm \frac{5}{31} \times 100\% = 16\%$

(d) The percentage uncertainty in measuring the temperature of the water is much less than the uncertainty in measuring the time. (±1 °C, leading to ±1 to 2%).

This example shows the reasoning in estimating the uncertainty in a measured quantity and how to calculate percentage uncertainty. You might feel that 5 s is rather a large uncertainty in measuring the time. It is at the upper limit, and you might be justified in claiming the uncertainty to be as little as 1 s. Nevertheless, if you try the experiment for yourself, and repeat it two or three times (as you should do with something this subjective), you will find that an uncertainty of 5 s is not unreasonable. The measurement of the initial temperature of the water has a much lower percentage uncertainty as less judgement is needed to make the measurement.

The next stage is to look at how to test whether a hypothesis is justified or not.

Worked example 2

It is suggested that the time taken (t) to melt an ice cube is inversely proportional to θ^2, where θ is the initial temperature of the water in °C.

Explain whether or not your results from Worked example 1 support this theory.

Answer

If $t \propto 1/\theta^2$ then $t \times \theta^2$ = constant

Trial 1: $85 \times 50^2 = 213\,000$
Trial 2: $31 \times 80^2 = 198\,400$

difference between the constants = 14 600
percentage difference = $(14\,600/198\,400) \times 100\% = 7.4\%$

This is less than the calculated uncertainty in the measurement of t (= 16%, for trial 2) so the hypothesis is supported.

There are various ways of tackling this type of problem — this is probably the simplest. Note that it is important to explain fully why the hypothesis is/is not supported. At the simplest level, if the difference between the two calculated values for the constant is greater than the percentage uncertainties in the measured quantities, then the evidence would not support the hypothesis.

Hint

A more sophisticated approach in this example would be to consider the combined uncertainties in the raw readings as the limit at which the experiment supports the theory. The theory predicts that $t = \text{constant}/\theta^2$, which means that the constant $= t \times \theta^2$. To combine uncertainties on multiplication (or division) the percentage uncertainties are added.
- percentage uncertainty in θ = 2%, (see above, the greatest uncertainty is chosen)
- therefore, percentage uncertainty in $\theta^2 = 2 \times 2\% = 4\%$,
- percentage uncertainty in t = 16%
- total uncertainty = 4% + 16% = 20%

Evaluating the experiment

There are two parts to this section:
- identifying weaknesses in the procedure
- suggesting improvements that would increase the reliability of the experiment

Before looking at Worked example 3, try to list *four* weaknesses in the procedure in the previous experiment. Then list *four* improvements that would increase the reliability of the experiment.

Worked example 3
State four sources of error or limitations of the procedure in Investigation 3 — the melting ice experiment.

Answer
(1) Two readings are not enough to make firm conclusions.
(2) The ice cubes may not be the same mass.
(3) There will be some energy exchanges with the surroundings.
(4) The ice cubes might be partly melted before they are put into the water.

Identifying weaknesses in a procedure is not easy but the more practical work you do the better you will become. It is important to be precise when making your points. In many experiments (not this one!) parallax can lead to errors. It would not be enough to say in an answer 'parallax errors', you would need to identify where those errors arose. If you were trying to measure the maximum amplitude of a pendulum, you would need to say, 'Parallax errors, when judging the highest point the pendulum bob reaches.'

Having identified the areas of weakness you now need to suggest how they could be rectified. The list given in Worked example 4 is not exhaustive — for example a suggestion that there should be the same volume of water in the beaker every time would also be sensible. However, a comment regarding measuring the average temperature of the water would not be acceptable as this would make it a different, albeit a perfectly valid, experiment.

If you have not got four weaknesses try writing 'cures' for the weaknesses suggested in Worked example 3.

Worked example 4
Suggest four improvements that could be made to Investigation 3. You may suggest the use of other apparatus or different procedures.

Answer
(1) Take more sets of readings with the water at different temperatures and plot a graph of t against $1/\theta^2$.
(2) Weigh the ice cubes.
(3) Carry out the experiment in a vacuum flask.
(4) Keep the ice cubes in a cold refrigerator until required.

In many ways this is easier than identifying weaknesses but note that you need to make clear what you are doing. The first suggestion is a good example — there is no point in taking more readings unless you do something with them! Note also the answer makes it clear that it is not just repeat readings that would be taken (that should have been done anyway); it is readings at different water temperatures.

This experiment does not cover all the difficulties you might encounter; for instance in the bouncing ball experiment (Investigation 1), the major difficulty is measuring the height to which the ball bounces. One possible way in which this problem could be solved is to film the experiment and play it back frame by frame or in slow motion.

Whenever you carry out an experiment, think about the weaknesses in the procedure and how you would rectify them. Discuss your ideas with your friends and with your teacher. You will find that you gradually learn the art of critical thinking, which will help you to score highly on this part of the paper.

AS
Questions
&
Answers

AS Questions & Answers

This section provides a practice examination paper similar to Paper 2. All the questions are based on the topic areas described in the previous sections of this book.

You have 1 hour to complete the paper. There are 60 marks on the paper so you can spend one minute per mark. If you find that you are spending too long on one question, move on to another that you can answer more quickly. If you have time at the end, then come back to the difficult one.

Not only will you have to use your knowledge to answer the questions, you must also target each answer to the question. There are two hints to tell you how much you need to write: the space available for the answer and the number of marks. The latter is the better hint. The more marks there are for a question the more points you have to make. As a rough guide, in an explanation or description, 1 mark is awarded for each relevant point made.

Look carefully at exactly what each question wants you to do. For example, if a question asks you to 'State' what happens then you only need to tell the examiner what would happen, without any further explanation. You will not lose marks for giving extra information unless you contradict yourself, but neither will you gain extra credit — all you do is waste precious time. On the other hand, if the question asks you to 'Explain' then you need to say how or why something happens, not just describe what happens. Otherwise, you will lose marks.

At AS a lot of the marks are for quantitative work. In many ways these are easy marks to gain. However, many students throw marks away unnecessarily. You should show each step in your working, and where necessary explain what you are doing. If you make an error and the examiner can see where you have gone wrong, credit may still be given. If the examiner cannot see where you have made a mistake, then he or she cannot give any credit. Working logically, step by step, will help you to see what you are doing and will reduce your chances of making errors. It might take a moment or two longer but the rewards are worth it.

Exemplar paper
Question 1

The frequency f of a stationary wave on a string is given by the formula:

$$f = \frac{T}{4L^2 m}$$

where L is the length of the string, T is the tension in the string, m is the mass per unit length of the string and k is a dimensionless constant.

(a) State which of the quantities, f, L, T and m are base quantities. (1 mark)

(b) (i) State the units of f, L, T and m in terms of base units. (3 marks)

   (ii) By considering the homogeneity of the equation, determine the value of k. (2 marks)

Total: 6 marks

Candidate A
(a) T and m ✗

   ✎ The candidate does not understand the term base quantity. He or she seems to think that it means a quantity that is not raised to a power. 0/1

(b) (i) f is in hertz ✗, L is in metres ✓, T is in newtons ✗ and m is in kg m^{-1} ✓.

   ✎ Again the candidate shows a lack of understanding of base units. Hertz is a derived unit, the unit of L is correct, the newton is a derived unit — kg m s^{-2}, the answer for m (mass per unit length) is correct. Note that there are four responses required for 3 marks. The candidate has two correct responses and earns 1 mark. 1/3

   (ii) units of left-hand side of the equation are $(1/s)^k$

   units of right-hand side of the equation are N/m^2 kg m^{-1}
   = k̶g̶ m̶ s^{-2}/m̶^{2} k̶g̶ m̶$^{-1}$ ✓

   $(1/s)^k = s^{-2}$, so $k = -2$ ✗

   ✎ Despite difficulty with base quantities this is done quite well, although it is a shame that the candidate has muddled a fraction and the negative index in the last line. (1/2)

Candidate B
(a) L only ✓

   ✎ L is the only base unit so the candidate scores this mark.

(b) (i) f is in s^{-1} ✓, L is in metres ✓, T is kg m s^{-2} ✓ and m is in kg m^{-1} ✓

   ✎ All four are correct, for the full 3 marks. 3/3

119

AS Questions & Answers

(ii) units of left-hand side of the equation are $(1/s)^k$

units of right-hand side of the equation are $N/m^2 \, kg \, m^{-1}$ = k̶g̶m̶s^{-2}/m̶k̶g̶ ✓

$s^{-k} = s^{-2}$, so $k = 2$ ✓

📝 All correct. 2/2

Question 2

A builder throws a brick up to a second builder on a scaffold, who catches it. The graph shows the velocity of the brick from when it leaves the hand of the first builder to when the second builder catches it.

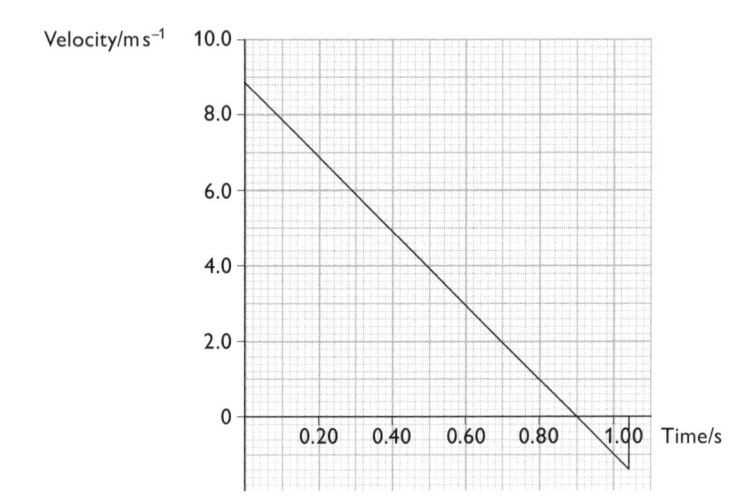

(a) Show that the acceleration is $9.8 \, m \, s^{-2}$. (2 marks)

(b) The gradient of the velocity–time graph is negative.

Explain what this shows. (1 mark)

(c) The second builder catches the brick 1.04 s after the first builder released it.

Calculate the height the second builder is above the first builder. (2 marks)

(d) The second builder drops a brick for the builder on the ground to catch.

Suggest why it is much more difficult to catch this brick than the one in the previous case. (1 mark)

Total: 6 marks

Candidate A

(a) acceleration = gradient of the graph = 8.8 − 0/(0.90 − 0)

= 9.8 m s⁻² ✓ ✗

> The candidate recognises that the accleration is equal to the gradient, and correctly identifies suitable points on the graph. However, the lower line of the equation should read (0 − 0.9). A compensation mark is given.

(b) The brick is slowing down ✓.

> This is not a very convincing statement but it is just enough to earn the mark. 1/1

(c) distance = area under the graph

= (½ × 8.8 × 0.90) + ✗ (½ × 1.40 × 0.14) ✓

= 4.1 m

> The candidate recognises that the area under the graph is equal to the distance travelled but unfortunately does not realise that for the last 0.14 s the brick is moving downwards, so the velocity is negative. Nevertheless, it is easy to spot the error so only 1 mark is lost. 1/2

(d) The brick will be moving downwards not upwards ✗.

> It is not true! The brick had started to move downwards in the earlier example. Neither does it answer the question. 0/1

Candidate B

(a) acceleration = gradient of the graph = 8.8 − 0/(0 − 0.90)

= −9.8 m s⁻² ✓ ✓

> All correct. 2/2

(b) The velocity and acceleration are in opposite directions. This shows deceleration ✓.

> This is a much more convincing answer than that of Candidate A. 1/1

(c) distance = area under the graph

= (½ × 8.8 × 0.90) − (½ × 1.40 × 0.14) ✓

= 3.9 m ✓

> All correct. The student has sensibly rounded the answer to two significant figures. (2/2)

(d) The brick will be travelling much faster. In the earlier example it is almost stationary when it reaches the builder on the scaffold.

> This is a good answer. 1/1

Question 3

A glider on an air track has a mass 1.2 kg. It moves at 6.0 m s^{-1} towards a second stationary glider of mass 4.8 kg. The two gliders collide, and the incoming glider rebounds with a speed of 3.6 m s^{-1}.

(a) Show that the speed of the second glider after the collision is 2.4 m s^{-1}. (3 marks)

(b) Show that the collision is elastic. (3 marks)

The gliders are in contact for 30 ms during the collision.

(c) (i) Calculate the average force on the stationary glider during the collision. (2 marks)

(ii) Compare the forces on the two gliders during the collision. (2 marks)

Total: 10 marks

Candidate A
(a) momentum before the collision = 1.2 × 6.0 = 7.2 kg m s^{-1} ✓

momentum after the collision = (4.8 × 2.4) + (1.2 × 3.6) ✗

= 11.52 + 4.32 = 15.84 kg m s^{-1} ✗

- The candidate earns 1 mark for the correct calculation of the initial momentum. However, there is a failure to recognise that momentum is a vector and direction must be included. 1/3

(b) K.E. before = ½mv² = ½ × 1.2 × 6.0 = 3.6 J ✗

K.E. after = ½ × 4.8 × 2.4² + ½ × 1.2 × 3.6² = 13.824 + 7.776 = 21.6 J ✓

K.E. before the collision = K.E. after the collision? ✓

- The candidate writes down the correct formula for kinetic energy but forgets to square the velocity when calculating the K.E. before the collision. The remainder of the calculation is fine. The final mark is for recognising that the kinetic energy before the collision should be equal to the kinetic energy after the collision. It is even recognised (by the question mark) that something has gone wrong. 2/3

(c) (i) acceleration of glider = 2.4/(30 × 10^{-3}) = 80 m s^{-2}

F = ma = 4.8 × 80 = 384 ✓ ✗

(ii) Force is smaller than the other ✗ because the glider is smaller and it is in the opposite direction ✓. Equal to the force on the other glider.

- The answer in part (i) is a valid way of calculating the force but this is a case where the candidate is expected to provide the unit since the answer cue does not provide one. In part (ii) there is recognition that the forces are opposite in direction but not that they are of the same magnitude. Indeed, there is a contradiction in the answer — the candidate first says that the force is smaller and then that it is equal to the force on the other glider. 2/4

Candidate B
(a) momentum before the collision = 1.2 × 6.0 = 7.2 kg m s⁻¹ ✓

momentum after the collision = (4.8 × 2.4) + (1.2 × −3.6) ✓

= 11.52 − 4.32 = 7.2 kg m s⁻¹ = momentum before the collision ✓

☻ All correct. 3/3

(b) In elastic collisions kinetic energy is conserved ✓.

K.E. before = ½ × 1.2 × 6.0² = 21.6 J ✓

K.E. after = ½ × 4.8 × 2.4² + ½ × 1.2 × 3.6² = 13.824 + 7.776 = 21.6 J ✓

☻ All correct. 3/3

(c) (i) force = rate of change of momentum = 4.8 × (2.4 − 0)/(30 × 10⁻³) ✓

= 384 N ✓

(ii) Force is the same magnitude ✓ but in the opposite direction ✓

☻ All correct. 4/4

Question 4

(a) (i) Explain what is meant by the *moment of a force* about a point. (2 marks)

The diagram shows the principle of a hydraulic jack. A vertical force is applied at one end of the lever which is pivoted at A. The plunger is pushed down creating a pressure on the oil, which is pushed out of the master cylinder, through the valve into slave cylinder.

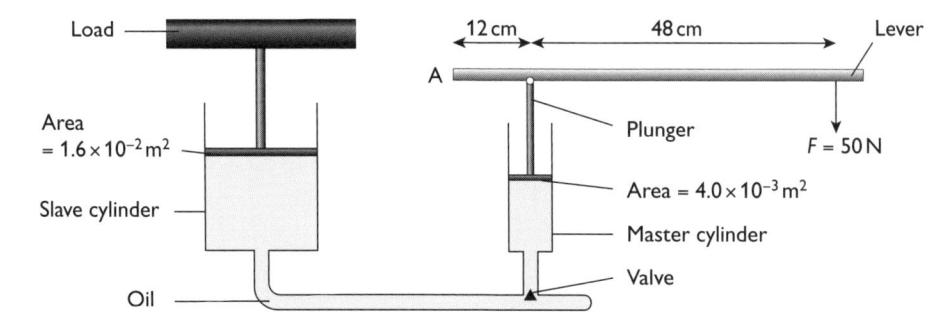

Load —
Area = 1.6 × 10⁻² m²
Slave cylinder
Oil
A
12 cm   48 cm   Lever
Plunger
F = 50 N
Area = 4.0 × 10⁻³ m²
Master cylinder
Valve

(b) (i) Calculate the force produced on the plunger by the lever. (2 marks)

(ii) Calculate the pressure exerted on the oil by the plunger. (2 marks)

AS Questions & Answers

The pressure is transmitted through the oil so that the same pressure is exerted in the slave cylinder.

(c) Calculate the load the jack can support. (1 mark)

(d) Suggest two design changes to the jack so that a larger load could be lifted. (2 marks)

Total: 9 marks

Candidate A

(a) moment of a force = force × distance from the pivot ✗ ✗

> The candidate has some idea of the concept of moment but the explanation is poor. There is no reference to the distance being the perpendicular distance from force to the pivot. 0/2

(b) (i) 50 × 48 ✗ = 12 F

$F = 200\,\text{N}$ ✗

(ii) pressure = force/area = $200/(4 \times 10^{-3})$ ✓

= 50 000 Pa ✓

> The candidate has not clarified the point that the moments are being taken about. As a result, the distance between the load force and the pivot is wrong. The rest of the calculation is correct. The earlier error is carried forward so both marks for part (ii) are scored. 2/4

(c) load = pressure × area = $50\,000 \times 1.6 \times 10^{-2} = 800\,\text{N}$ ✓

> The error is carried forward again, so the mark is scored. 1/1

(d) (i) Make the handle of the lever longer ✓.

(ii) Increase the cross-sectional area of the cylinder ✗.

> Part (i) is correct. However, in part (ii) the candidate does not specify which cylinder's cross-sectional area should be increased. Increasing the area of the master cylinder would have the opposite effect to that required. 1/2

Candidate B

(a) The moment of a force is the magnitude of the force × the perpendicular distance of the force from the point ✓ ✓

> Correct. 2/2

(b) (i) Take moments about A:

50 × 60 = 12 F ✓

$F = 250\,\text{N}$ ✓

(ii) pressure = force/area = $250/(4 \times 10^{-3})$ ✓

= 62 500 Pa ✓

International AS and A Level Physics Revision Guide

   ☒ All correct. 4/4

(c) load = pressure × area = 62 500 × 1.6 × 10^{-2} = 1000 N ✓

   ☒ Correct. 1/1

(d) (i) Increase the length of the lever ✓.

   (ii) Increase the cross-sectional area of the slave cylinder ✓.

   ☒ Both correct. (2/2)

Question 5

(a) (i) Explain what is meant when two sources of light are described as *coherent* and state the conditions necessary for coherence. (2 marks)

   (ii) Explain why two separate light sources cannot be used to demonstrate interference of light. (2 marks)

(b) A Young's slits experiment is set up to measure the wavelength of red light.

   The slit separation is 1.2 mm and the screen is 3.0 m from the slits. The diagram shows the interference pattern that is observed.

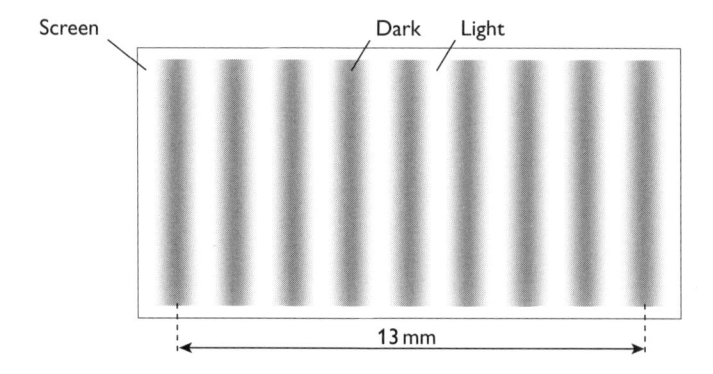

   Calculate the wavelength of the light. (2 marks)

(c) Explain how you would expect the pattern to change if the red light is replaced by a blue light. (2 marks)

Total: 8 marks

Candidate A

(a) (i) Two sources are coherent if they have no phase difference ✗ ✗.

   ☒ This is a common error. Students often ignore the fact that light from two sources can be coherent if there is a phase difference, provided that the phase difference is constant. The necessity for the two sources to have the same frequency is not mentioned. (0/2)

(ii) They are not coherent ✓.

✎ The statement is correct and so gains a mark. However, the question asks the candidate to *explain* and there is no explanation given as to why the sources are not coherent. 1/2

(b) $n\lambda = ax/D \rightarrow 9\lambda$ ✗ $= 1.2 \times 10^{-3} \times 13 \times 10^{-3}/3\lambda = 5.8 \times 10^{-7}$ m ✓ (e.c.f.)

✎ The candidate has counted the nine minima but there are only eight fringes. The rest of the calculation is completed correctly, for 1 mark. (1/2)

(c) Blue light has a longer wavelength than red light ✗ so the fringes would be further apart ✓.

✎ The candidate should be aware that the blue end of the spectrum has the shortest wavelengths of visible light. The correct conclusion has been drawn from the original error, so the second mark is scored. 1/2

Candidate B

(a)(i) Two sources are coherent if they have a constant phase difference ✓. For this, they must have the same frequency ✓.

✎ Correct. 2/2

(ii) Light is not emitted as a single wave train but as short wave trains. The phases of the different wave trains are random ✓, so the wave trains from the two sources are not coherent ✓.

✎ Correct. 2/2

(b) $n\lambda = ax/D \rightarrow 8\lambda = 1.2 \times 10^{-3} \times 13 \times 10^{-3}/3$ ✓

$\lambda = 6.5 \times 10^{-7}$ m ✓

✎ Correct. 2/2

(c) Blue light has a shorter wavelength than red light ✓ so the fringes would be closer together ✓.

✎ This is a good explanation. The candidate has noted that a reason is needed. 2/2

Question 6

(a) Explain what is meant by the *electric field strength* at a point. (1 mark)

The diagram shows two parallel plates in an evacuated tube. The earthed plate is heated so as to emit electrons. The plates are 8.0 cm apart, with a potential difference of 2.0 kV across them.

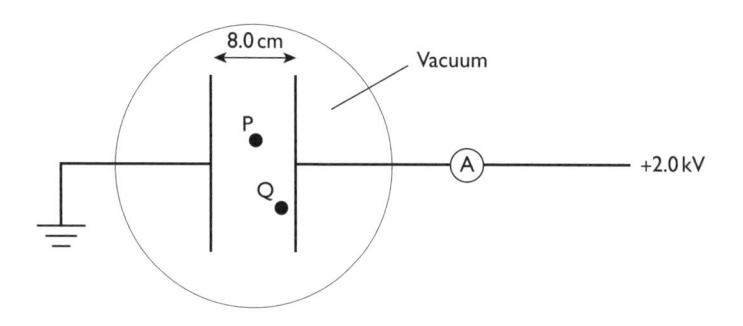

International AS and A Level Physics Revision Guide

(b) Calculate the electric field strength between the two plates. (2 marks)

(c) (i) An electron is at point P midway between the two plates. Calculate the force on it. (2 marks)

(ii) A second electron is at point Q, near the positive plate. State the force on this electron, giving a reason for your answer. (2 marks)

(d) A current of 2.4 μA is recorded on the ammeter.

Calculate the number of electrons that move across from one plate to the other in 1 minute. (2 marks)

Total: **9 marks**

Candidate A

(a) Electric field strength is the force per unit charge ✗.

> The candidate has missed that it should be the force on a positive charge and has also not emphasised that the force is on a point charge placed at that point.

(b) $E = V/d = 2000/(8 \times 10^{-2})$ ✓

$= 25000 \, \text{NC}^{-1}$ ✓

> Correct. NC^{-1} is an alternative unit for electric field strength. 2/2

(c) (i) $E = F/Q \rightarrow 25000 = F/(1.6 \times 10^{-19})$ ✓

$F = 4 \times 10^{-15} \, \text{N}$ ✓

(ii) Greater than $4.0 \times 10^{-15} \, \text{N}$ ✗. The force is stronger because the charge is nearer the positive plate ✗.

> Part (i) is correct. However, the response to part (ii) shows that the candidate does not understand the concept of a uniform field. 2/4

(d) $Q = It, = 2.4 \times 10^{-6} \times 60 = 1.44 \times 10^{-4} \, \text{C}$ ✓

> The candidate has correctly calculated the charge that has passed but does not know how to proceed. 1/2

AS Questions & Answers

Candidate B

(a) Electric field strength at a point is the force per unit positive charge acting on a stationary point charge at placed at that point ✓.

🖉 This is an excellent definition. 1/1

(b) $E = V/d = 2000/(8 \times 10^{-2})$ ✓

$= 25000 \, \text{V m}^{-1}$ ✓

🖉 Correct. 2/2

(c) (i) $E = F/Q \rightarrow 25000 = F/(1.6 \times 10^{-19})$ ✓

$F = 4.0 \times 10^{-15} \, \text{N}$ ✓

(ii) $4.0 \times 10^{-15} \, \text{N}$ ✓. The field is uniform, therefore the force on the charge is the same anywhere between the plates ✓.

🖉 All correct. 4/4

(d) $Q = It$

number of electrons × charge on an electron = Q

so $ne = It$ ✓

$n = 2.4 \times 10^{-6} \times 60/(1.6 \times 10^{-19}) = 9.0 \times 10^{14}$ ✓

🖉 All correct. 2/2

Question 7

(a) Explain the difference between the terminal potential difference and the e.m.f. of a cell. (2 marks)

(b) A student makes a 4.0 Ω resistor by winding 3.9 m of insulated eureka wire of diameter 0.78 mm around a wooden former.

Calculate the resistivity of eureka. (2 marks)

(c) When the student connects the resistor across the terminals of a cell of e.m.f. 1.56 V there is a current of 0.37 A.

Calculate the internal resistance of the cell. You may assume that the ammeter has negligible resistance. (2 marks)

Total: 6 marks

Candidate A

(a) The terminal potential difference is numerically equal to the electrical energy converted to other forms of energy when it moves around the circuit from one terminal to the other ✓. The e.m.f. is the terminal potential difference when no current flows ✗.

🅔 The first part is answered very well. The second answer, although a good rule of thumb, is not how e.m.f. is defined (see Candidate B's answer). 1/2

(b) $R = \rho L/A \rightarrow 4 = \rho \times 3.9/(\pi(0.78 \times 10^{-3})^2)$ ✗

$\rho = 1.96 \times 10^{-6}\,\Omega\text{m}$ (e.c.f.) ✓

🅔 This is a good attempt, except that the candidate has failed to divide the diameter by two. Cross-sectional area = πr^2. 1/2

(c) $V = IR = 0.37 \times 4 = 1.48$

lost volts = $1.56 - 1.48 = 0.08\,\text{V}$ ✓

🅔 This is a valid way of doing the problem but after a good start the candidate does not know how to complete the problem — internal resistance = lost volts/current = $0.08/0.37 = 0.22\,\Omega$. 1/2

Candidate B

(a) The terminal potential difference is numerically equal to the electrical energy converted to other forms of energy when unit charge moves around the circuit from one terminal to the other ✓. The e.m.f. is numerically equal to the energy given to unit charge when it passes through the cell ✓.

🅔 An excellent answer. 2/2

(b) $R = \rho L/A \rightarrow 4 = \rho \times 3.9/(\pi(0.78 \times 10^{-3}/2)^2)$ ✓

$\rho = 4.9 \times 10^{-7}\,\Omega\text{m}$ ✓

🅔 Correct. 2/2

(c) $E = IR + Ir \rightarrow 1.56 = (0.37 \times 4) + 0.37r$ ✓

$r = 0.22\,\Omega$ ✓

🅔 Correct. 2/2

Question 8

In a nuclear reaction, a neutron collides with the osmium nuclide, $^{190}_{76}\text{Os}$.

(a) Deduce the number of protons and neutrons in the osmium nuclide. (1 mark)

(b) The neutron is absorbed by the nucleus, which then undergoes beta decay, to form an iridium (Ir) nucleus.

Write an equation to describe this reaction. (2 marks)

(c) The mass defect per nucleon of the daughter nuclide, iridium, is slightly greater than the mass defect per nucleon of the osmium nuclide.

Explain what information this gives about the two nuclides. (3 marks)

Total: 6 marks

Candidate A
(a) number of protons = 76, number of neutrons = 190 ✗

> The number of protons is correct but the top number is the number of nucleons (protons + neutrons), not the number of neutrons. 0/1

(b) $^{190}_{76}\text{Os} + ^{1}_{0}\text{n} \rightarrow ^{191}_{75}\text{Ir} + ^{0}_{-1}\beta$ ✓ ✗

> The nucleon numbers add up correctly showing the conservation of nucleon numbers. However, the candidate has not considered that the proton number of the beta particle is −1. 1/2

(c) Greater mass defect means there is more binding energy ✓, so the iridium nucleus has more energy, some of the mass has been turned into this binding energy ✗ ✗.

> The answer starts off quite well with a correct initial statement. However, the candidate does not understand binding energy. It is the energy that must be put in to completely separate the nucleons, meaning that the nucleus is in a lower energy state and is therefore more stable. The nucleus has less mass because it has less energy (it is in a lower energy state) than before. 1/3

Candidate B
(a) number of protons = 76, number of neutrons = 114 ✓

> Both correct. 1/1

(b) $^{190}_{76}\text{Os} + ^{1}_{0}\text{n} \rightarrow ^{191}_{77}\text{Ir} + ^{0}_{-1}\beta$ ✓ ✓

> The candidate has not included the intermediate stage of the formation of a nuclide with nucleon number 191 and proton number 76. Nevertheless, the starting and finishing points are correct and the answer deserves full credit. 2/2

(c) Greater mass defect means the nucleus is in a lower potential energy state ✓, so more energy is required to tear it apart into its constituent parts ✓. It is, therefore, more stable than the osmium nucleus ✓.

> This is a good answer despite the non-scientific term, 'tear it apart'. To have written 'to completely separate the nucleons' would be better. The binding energy is not mentioned but an understanding of its meaning is shown in the second sentence. 3/3.

A2 Content Guidance

General physics
Physical quantities and units

There is very little new syllabus content in this section. The **mole** was referred to in the list of SI units. This work is discussed in detail in the section on ideal gases (pages 144–148).

Measurement techniques

The only new measuring device introduced at A2 is the **Hall probe**. The use of this is described in the section on magnetic fields (page 192).

Newtonian mechanics
Motion in a circle

Radian measurement

You are familiar with the use of degrees to measure angles, with a complete circle equal to 360°. There is no real reason why a circle is split into 360° — it probably arises from the approximate number of days it takes for the Earth to orbit the Sun.

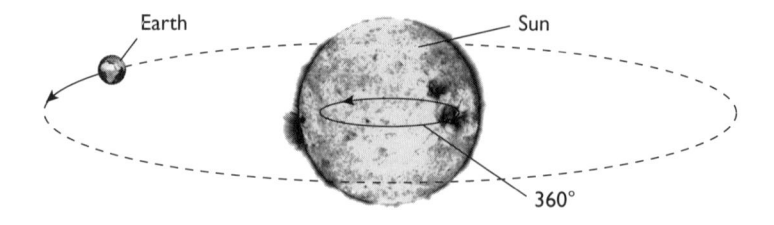

It is much more convenient to use radians, where the angle in radians is the ratio of the arc length to the radius:

angle (in radians) = arc length/radius

This means that one radian is the angle subtended at the centre of a circle by an arc length equal in length to the radius.

The circumference of a circle = $2\pi r$, where r is the radius. Hence, the angle subtended by a complete circle (360°) = $2\pi r/r = 2\pi$.

360° = 2π radians. This can be expressed as 1° = $2\pi/360$ radians or 1 radian = $(360/2\pi)°$.

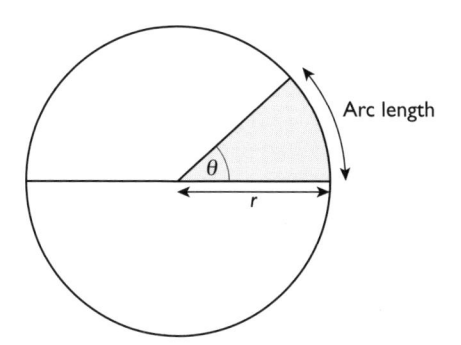

Worked example
(a) Convert the following angles to radians:
   (i) 180°
   (ii) 90°
   (iii) 60°
   (iv) 45°

(b) Convert the following angles to degrees:
   (i) $\pi/4$ rad
   (ii) $2\pi/3$ rad
   (iii) 1.0 rad

Answer
(a) (i) $180° = 180 \times 2\pi/360 = \pi$ rad
   (ii) $90° = 90 \times 2\pi/360 = \pi/2$ rad
   (iii) $60° = 60 \times 2\pi/360 = \pi/3$ rad
   (iv) $45° = 45 \times 2\pi/360 = \pi/4$ rad

(b) (i) $\pi/4$ rad $= \pi/4 \times 360/2\pi = 45°$
   (ii) $2\pi/3$ rad $= 2\pi/3 \times 360/2\pi = 120°$
   (iii) 1.0 rad $= 1.0 \times 360/2\pi = 57.3°$

Angular displacement and angular velocity
Consider a particle moving at constant speed (v) round a circle.

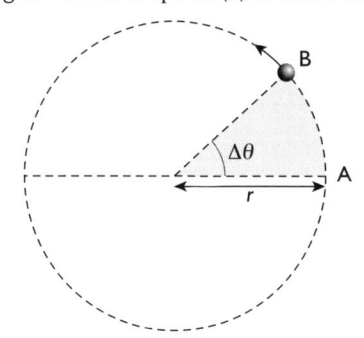

Angular displacement is defined as the change in angle (measured in radians).

As the particle moves round the circle, the angular displacement increases at a steady rate. The rate of change in angular displacement is called the angular velocity (ω). Angular velocity is therefore defined as the change in angular displacement per unit time:

$\omega = \Delta\theta/\Delta t$

Comparison with translational motion

Many of the concepts we met in kinematics at AS have their equivalent in circular motion. This is shown in Table 14.

Table 14

Translational motion			Circular motion		
Quantity	Unit	Relationships	Quantity	Unit	Relationships
Displacement (s)	m		Angular displacement (θ)	rad	
Velocity (v)	m s^{-1}	$v = \Delta s/\Delta t$	Angular velocity (ω)	rad s^{-1}	$\omega = \Delta\theta/\Delta t$

Look at the diagram at the bottom of page 133.

$\omega = \Delta\theta/\Delta t$

but $\Delta\theta = AB/r$

Therefore $\omega = AB/r\Delta t$

$AB/\Delta t$ = distance travelled/time = v

thus $\omega = v/r$ or, rearranging the formula, **$v = \omega r$**

> **Worked example**
> A car is travelling round a circular bend of radius 24 m at a constant speed of 15 m s^{-1}.
>
> Calculate the angular velocity of the car.
>
> **Answer**
> $\omega = v/r = 15/24 = 0.625 \approx 0.63$ rad s^{-1}

Constant speed, constant acceleration

We have already seen how a body can move at constant speed around a circle, but what is meant when it is said that the body has a constant acceleration?

To understand this you must remember the definition of acceleration: the change in velocity per unit time. Velocity, unlike speed, is a vector and so a change in direction is an acceleration.

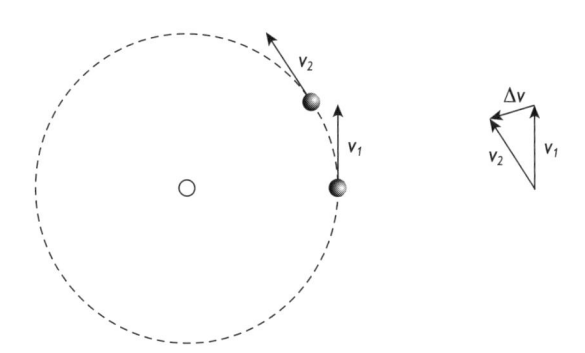

Consider a particle moving round a circle. At time t it has a velocity of v_1. After a short interval of time, Δt, it has the velocity v_2 — the same magnitude, but the direction has changed. The vector diagram shows the change of velocity Δv. You can see that this is towards the centre of the circle, the acceleration being $\Delta v/\Delta t$. As the body moves round the circle, the direction of its velocity is continuously changing, the change always being towards the centre of the circle. Thus the particle has an acceleration of constant magnitude but whose direction is always towards the centre of the circle. Such an acceleration is called a **centripetal acceleration**.

The magnitude of the acceleration is given by the formulae:

$$a = \frac{v^2}{r} \text{ and } \boldsymbol{a} = \omega^2 \boldsymbol{r}$$

Centripetal force and acceleration

It is important to realise that a body travelling round a circle at constant speed is *not* in equilibrium. From Newton's laws you will remember that for a body to accelerate a resultant force must act on it. The force must be in the same direction as the acceleration. Hence the force is always at right angles to the velocity of the body, towards the centre of the circle (a centripetal force). Such a force has no effect on the magnitude of the velocity; it simply changes its direction.

Using the relationship $F = ma$ (where F = force and m = mass of the body) we can see that the force can be calculated from the formulae:

$$F = \frac{mv^2}{r} \text{ and } \boldsymbol{F} = \boldsymbol{m\omega^2 r}$$

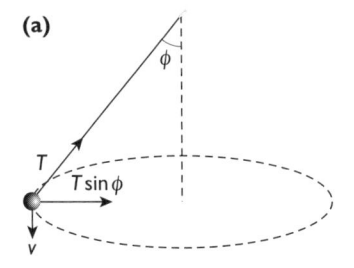

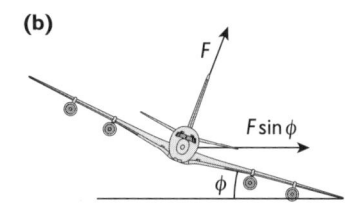

II Newtonian mechanics

Diagram (a) above shows a rubber bung being whirled round on a string. The string is under tension. The centripetal force is the component of the tension in the horizontal direction ($T\sin\phi$).

$$\frac{mv^2}{r} = T\sin\phi$$

In diagram (b) the uplift on the aeroplane is perpendicular to the wings. When the aeroplane banks there is a horizontal component to this, which provides a centripetal force ($F\sin\phi$) and the plane moves along the arc of a circle.

$$\frac{mv^2}{r} = F\sin\phi$$

Worked example

The diagram shows a racing car rounding a bend of radius 120 m on a banked track travelling at 32 m s^{-1}.

(a) Calculate the angle ϕ if there is no tendency for the car to move either up or down the track. You may treat the car as a point object.

(b) Suggest and explain what would happen if the car's speed was reduced.

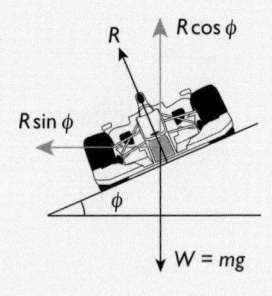

Answer

(a) R is the normal reaction force.

Resolving vertically: $R\cos\phi = mg$

Resolving horizontally: $R\sin\phi = mv^2/r$

Dividing the two equations:

$\sin\phi/\cos\phi = (mv^2/r)/mg$

$\tan\phi = v^2/gr = 32^2/(9.8 \times 120) = 0.871$

$\phi = 41°$

(b) The car would tend to slip down the slope as the required centripetal force would be less. In practice, frictional forces would probably mean that it would continue in a circle of the same radius.

> **Hint**
>
> The car is clearly not a point object but modelling it as one simplifies the problem. The normal reaction is, in reality, shared at each of the four wheels. The wheels on the outside of the curve travel in a larger circle than those on the inside of the circle. Engineers and scientists often use simplified models, which they then develop to solve more complex problems.

Gravitational field

A gravitational field is a region around a body that has mass, in which another body with mass experiences a force.

The uniform field

Any object near the Earth's surface is attracted towards the Earth with a force that is dependent on the mass of the body. Similarly, an object near the Moon is attracted towards the Moon's surface but now the force is smaller. The reason for this is that the **gravitational field strength** is greater near the Earth than it is near the Moon.

force = mass × gravitational field strength

In symbols:

$F = mg$

You might remember g as the acceleration due to gravity or acceleration of free fall but if you compare the formulae $F = ma$ and $F = mg$, you can see that the acceleration due to gravity and the gravitational field strength are the same thing.

Gravitational field strength at a point is defined as the gravitational force per unit mass at that point.

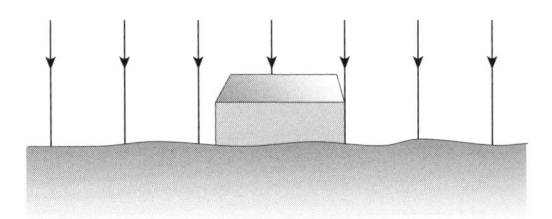

The gravitational field strength near the Earth's surface is uniform

Gravitational forces between point objects

It is not just large objects that attract each other — all masses have a gravitational field. This means that they attract other masses.

II Newtonian mechanics

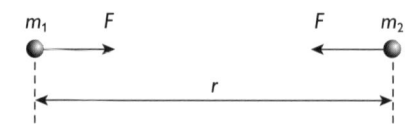

Two point masses of mass m_1 and m_2 separated by a distance r will attract each other with a force given by the formula:

$$F = -\frac{Gm_1m_2}{r^2}$$

where G is a constant known as the **universal gravitational constant**. Its value is $6.67 \times 10^{-11}\,N\,m^2\,kg^{-2}$.

This is known as Newton's law of gravitation.

The minus sign in the equation shows the vector nature of the force. It is often easier to consider only the magnitude of the force, hence the minus sign may be omitted. This practice will be adopted in this book.

It is often useful, particularly when considering planet-sized objects, to write the equation as:

$$F = \frac{GMm}{r^2}$$

Although slightly more complex with bodies of finite size, all the mass of any object can be considered to act at a single point, which is called **the centre of mass**. This simplifies the maths and in effect the object is treated as a point mass. However, we must be careful to remember to measure any distances between objects as the distance between their centres of mass. Note that the formula above assumes that planets can be treated as point objects.

Worked example

Two spheres of radius 0.50 cm and masses 150 g and 350 g are placed so that their centres are 4.8 cm apart.

(a) Calculate the force on the 150 g sphere.

(b) Write down the force on the 350 g sphere.

Answer

(a) 350 g = 0.35 kg, 150 g = 0.015 kg, 4.8 cm = 0.048 m

$$F = \frac{Gm_1m_2}{r^2} = \frac{6.67 \times 10^{-11} \times 0.35 \times 0.15}{0.048^2} = 1.5 \times 10^{-9}\,N$$

(b) In accordance with Newton's third law, the force on the 350 g mass will also be $1.5 \times 10^{-9}\,N$ but in the opposite direction.

> **Hint**
>
> This shows how small the gravitational attraction between two small objects is. It is only when we consider planet-sized objects that the forces become significant.

Gravitational fields

The gravitational field strength has already been defined as the gravitational force per unit mass at that point.

At AS, you only considered gravitational fields on large objects such as the Earth and other planets, and then only near their surfaces. Under these circumstances, the field may be considered uniform. However, the gravitational field of a point object is radial. This is also true for any body of finite size if we move a significant distance from the body. In the latter case, the radial field is centred on the centre of mass of the body.

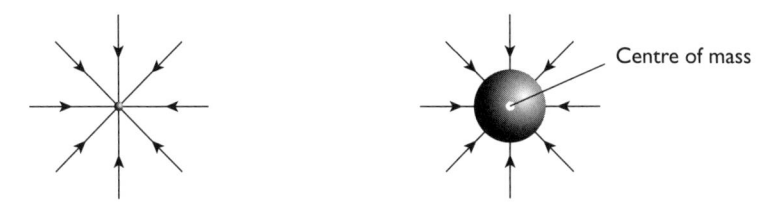

The gravitational field of a point mass and a body of finite size

You can see from the diagram above that the lines of gravitational force get further apart as the distance from the centre of mass increases. This shows that the field strength decreases with increasing distance from the body.

Consider the equation for the gravitational force between two objects and the definition of gravitational field strength:

$$F = \frac{GMm}{r^2} \text{ and } g = F/m$$

Substituting for F, the m cancels giving

$$g = \frac{GM}{r^2}$$

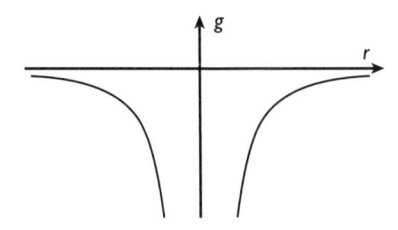

The gravitational field near a spherical body

The equation shows an **inverse square** relationship. This means if the distance from the mass is doubled the field decreases by a factor of 4 (2^2).

II Newtonian mechanics

Worked example
Calculate the gravitational field strength at the surface of Mars.

(radius of Mars = 3.4×10^3 km, mass of Mars = 6.4×10^{23} kg)

Answer

3.4×10^3 km = 3.4×10^6 m

$$g = \frac{F}{m}$$

Therefore $g = \frac{GM}{r^2} = \frac{6.67 \times 10^{-11} \times 6.4 \times 10^{23}}{(3.4 \times 10^6)^2} = 3.7\, \text{N kg}^{-1}$

The gravitational field strength is $3.7\,\text{N kg}^{-1}$ towards the centre of Mars.

Orbital mechanics

The diagram shows a satellite travelling in a circular orbit around the Earth.

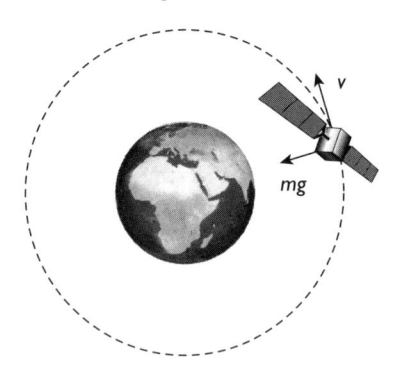

The gravitational pull on the satellite provides the centripetal force to keep the satellite in orbit.

Centripetal force:

$$F = \frac{mv^2}{r} = \frac{GMm}{r^2}$$

Cancelling m and r:

$v^2 = \frac{GM}{r}$ which can be rewritten as $v = \sqrt{\frac{GM}{r}}$

This can also be expressed in terms of angular velocity, ω:

$v = \omega r$, therefore $\omega = \sqrt{\frac{GM}{r^3}}$

You can see that the angular velocity, and hence the frequency and the period for one orbit, are dependent on the orbital radius.

The relation between the period T for one orbit and the angular velocity ω is:

$$T = \frac{2\pi}{\omega}$$

and that between the frequency f and the period is:

$$f = \frac{1}{T}$$

Worked example
A satellite is to be placed in a polar orbit 100 km above the Earth's surface.

Calculate
(a) the period of the orbit
(b) the speed of the satellite

(mass of Earth = 6.0×10^{24} kg, radius of Earth = 6.4×10^3 km)

Answer
(a) orbital radius of the satellite = Earth's radius + height of the satellite above the surface

$= (6.4 \times 10^3 + 100) = 6.5 \times 10^3$ km $= 6.5 \times 10^6$ m

$$\omega = \sqrt{\frac{GM}{r^3}} = \sqrt{\frac{6.67 \times 10^{-11} \times 6.0 \times 10^{24}}{(6.5 \times 10^6)^3}} = 1.2 \times 10^{-3} \text{ rad s}^{-1}$$

$$T = \frac{2\pi}{\omega} = \frac{2\pi}{1.2 \times 10^{-3}} = 5.2 \times 10^3 \text{ s} = 1.4 \text{ h}$$

(b) $v = \omega r$

orbital radius $= 6.5 \times 10^3$ km

$v = 1.2 \times 10^{-3} \times 6.5 \times 10^3 = 7.8$ km s^{-1}

Geostationary orbits
Imagine a satellite that is orbiting the Earth. Its orbital path is directly above the equator. If the satellite orbits in the same direction as the Earth spins and has an orbital period of 24 hours, it will remain over the same point above the Earth's surface. This type of orbit is used for communication satellites.

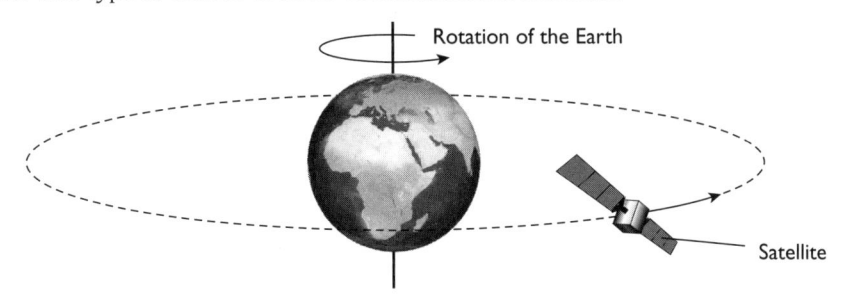

From the previous work you should be able to see that there is only one possible orbital radius for this type of satellite. With many countries requiring communications satellites, this means that a great deal of international cooperation is required.

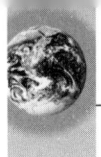

Worked example
Calculate the height above the Earth that a satellite must be placed for it to orbit in a geostationary manner.

(mass of Earth = 6.0 × 10²⁴ kg, radius of Earth = 6.4 × 10⁶ m)

Answer
time period required for a geostationary orbit is 24 h = 86 400 s

$$\omega = \frac{2\pi}{T}$$

$$\omega = \sqrt{\frac{GM}{r^3}}$$

Therefore: $\frac{2\pi}{T} = \sqrt{\frac{GM}{r^3}}$

and

$$r^3 = \frac{GMT^2}{(2\pi)^2} = \frac{6.67 \times 10^{-11} \times 6.0 \times 10^{24} \times 86400^2}{4\pi^2} = 7.57 \times 10^{22}$$

$r = \sqrt[3]{7.57 \times 10^{22}} = 4.23 \times 10^7$ m

This is the radius of the satellite's orbit. The radius of the Earth is 6.4 × 10⁶ m, so the height of the satellite above the Earth's surface is:

42.3 × 10⁶ − 6.4 × 10⁶ = 3.59 ×10⁷ m ≈ 3.6 × 10⁷ m

Gravitational potential

From earlier work you will be familiar with the idea that the gain in gravitational potential energy of a body when it is lifted through a height Δh is given by the formula:

$$\Delta W = mg\Delta h$$

This formula gives the change in gravitational potential energy. At what point does a body have zero potential energy? It is up to physicists to define the point at which a body has zero gravitational potential energy. It might seem sensible to choose the Earth's surface as this point. However, if we are considering work on the astronomical scale you can quickly see that this has no special significance. The point that is chosen is infinity — we say that the gravitational potential energy at an infinite distance from any other body is zero. This might seem a little difficult to start with; we know that a body loses potential energy as it approaches the Earth or other large body — therefore it has less than zero potential energy. This means that it has negative potential energy when it is near another body such as the Earth.

By considering the potential energy of a unit mass, we can assign each point in space a specific gravitational potential (Φ).

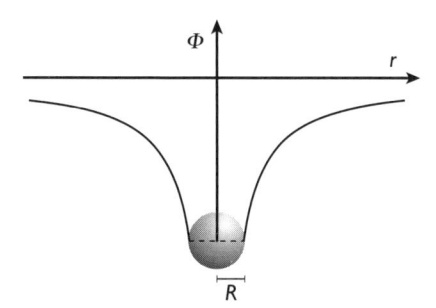

The gravitational potential near a body of radius R

The diagram above shows that the gravitational potential at the surface of the body is negative, and how the potential increases towards zero as we move away from the body.

Potential is defined as follows:

The potential at a point is the work done in bringing unit mass from infinity to that point.

Shape of the potential curve

A careful study of the potential curve shows it to be of the form $\Phi \propto 1/r$.

The formula for calculating the gravitational potential at a point is:

$$\Phi = -\frac{GM}{r}$$

where r is the distance from the centre of mass of the object.

Worked example

If a body is fired from the Earth's surface with sufficient speed, it can escape from the Earth's gravitational field.

(a) Calculate the potential at the Earth's surface.

(b) State and explain how much energy a body of unit mass would need to be given to escape from the Earth's field.

(c) Calculate the minimum speed at which the body must be fired to escape.

Answer

(a) $\Phi = -\dfrac{GM}{r} = -\dfrac{6.67 \times 10^{-11} \times 6.0 \times 10^{24}}{6.4 \times 10^{6}} = -6.25 \times 10^{7}\,\text{J}\,\text{kg}^{-1}$

(b) $6.25 \times 10^{7}\,\text{J}$, the energy required to reach infinity, zero potential energy.

(c) $E_k = \tfrac{1}{2}mv^2$, which leads to $v = \sqrt{\dfrac{2E_k}{m}} = \sqrt{\dfrac{2 \times 6.25 \times 10^{7}}{1}} = 1.1 \times 10^{4}\,\text{m}\,\text{s}^{-1}$

Matter

Ideal gases

The mole and the Avogadro constant

You are already familiar with the idea of measuring mass in kilograms and thinking of mass in terms of the amount of matter in a body. The mole measures the amount of matter from a different perspective — the number of particles in a body.

One mole is defined as the amount of substance that has the same number of particles as there are atoms in 12 g of carbon-12 isotope.

The amount of matter is a base quantity and the mole, consequently, is a base unit.

The abbreviation (unit) for the mole is **mol**.

The number of atoms in 12 g of carbon-12 is 6.02×10^{23}. This number is referred to as the Avogadro constant (N_A) and is written as 6.02×10^{23} mol^{-1}.

So:
- One mole of carbon-12 isotope contains 6.02×10^{23} carbon-12 atoms and has a mass of 12 g.
- One mole of helium-4 isotope contains 6.02×10^{23} helium-4 atoms and has a mass of 4 g.

Many gases are found not as single atoms but as diatomic molecules. For example, two hydrogen atoms form a H_2 molecule, so one mole of hydrogen contains 6.02×10^{23} hydrogen (H_2) molecules or 12.04×10^{23} atoms of hydrogen.

Worked example 1

Calculate the number of atoms in, and the mass of, the following:

(a)   1 mol of ozone (O_3)

(b)   3 mol of water (H_2O)

(relative atomic mass of oxygen = 16, relative atomic mass of hydrogen = 1)

Answer

(a)   1 mol of ozone = 6.02×10^{23} molecules = $3 \times 6.02 \times 10^{23}$ atoms
= 18.06×10^{23} atoms ≈ 18.1×10^{23} atoms

mass of ozone in 1 mol = $3 \times 16 = 48$ g

(b)   Each molecule of water contains 3 atoms (2 hydrogens, 1 oxygen).

number of atoms in 1 mol of water = $3 \times 6.02 \times 10^{23} = 18.06 \times 10^{23}$ atoms number of atoms in 3 mol = 3 × 18.06 × 10²³ ≈ 5.42 × 10²⁴ atoms

1 mol of water has mass = (2 × 1) + (1 × 16) = 18 g

Therefore the mass of 3 moles = 3 × 18 = 54 g

Worked example 2
The mass of 1 mol of hydrogen gas is 2 g. Calculate the mass of 1 hydrogen atom.

Answer
1 mol of hydrogen gas contains 2 × 6.02 × 10²³ hydrogen atoms

$$\text{mass of 1 hydrogen atom} = \frac{2}{2 \times 6.02 \times 10^{23}} = 1.66 \times 10^{-24} \text{ g}$$

The ideal gas equation

Experimental work shows that a fixed mass of any gas, at temperatures well above the temperature at which they condense to form liquids, and at a wide range of pressures, follow the relationships:
- at constant temperature — $p \propto 1/V$
- at constant pressure — $V \propto T$
- at constant volume — $p \propto T$

where p = pressure, V = the volume and T is the temperature measured on the Kelvin scale. The Kelvin scale of temperature is discussed further on page 151.

These three relationships can be combined to form a single equation:

$$\frac{pV}{T} = \text{constant}$$

This is more usually written as:

$pV = nRT$

where n is the number of moles of gas and R is the molar gas constant. The molar gas constant has the same value for all gases, 8.31 J K⁻¹. This equation is known as the equation of state for an ideal gas.

An ideal gas would follow this equation at all temperatures and pressures. Real gases, such as hydrogen, helium and oxygen, follow the equation at room temperature and pressure. However, if the temperature is greatly decreased or the pressure is very high they no longer behave in this way.

Worked example
Calculate the volume occupied by 48 mg of oxygen at 20°C and a pressure of 1.0 × 10⁵ Pa.

(relative atomic mass of oxygen = 16)

Answer
temperature = 273 + 20 = 293 K

oxygen forms diatomic O_2 molecules, so the mass of 1 mol of oxygen = 32 g

number of moles in 48 mg = $\dfrac{48 \times 10^{-3}}{32}$ = 1.5×10^{-3} mol

Using $pV = nRT$

$V = \dfrac{nRT}{P} = \dfrac{1.5 \times 10^{-3} \times 8.31 \times 293}{1.0 \times 10^{5}} = 3.7 \times 10^{-5}$ m^3

The kinetic theory of gases

Brownian motion
This was the defining evidence for the existence of moving molecules in fluids. It is described in detail on page 56, in the AS part of this book.

A model of a gas
You can now see that an ideal gas is modelled as consisting of many molecules that are moving randomly. The molecules themselves can be modelled as small unbreakable spheres that are spaced well apart and only interact with one another when they collide. If the gas is enclosed in a container, a pressure is produced on the container. This is due to the molecules colliding elastically with the container walls and the change in momentum of the particles producing a force on wall. It is the sum of all these forces that produces the pressure.

Relationship between molecule speed and pressure exerted
To show the relationship between the speed of the molecules in a gas and the pressure it exerts, the following assumptions are made:
- The forces between molecules are negligible (except during collisions).
- The volume of the molecules is negligible compared with the total volume occupied by the gas.
- All collisions between the molecules and between the molecules and the container walls are perfectly elastic.
- The time spent in colliding is negligible compared with the time between collisions.
- There are many identical molecules that move at random.

> **Note**
>
> It is worth noting that these assumptions effectively describe an ideal gas.

Consider a gas molecule of mass m in a cubic box of side L travelling at speed c parallel to the base of the box.

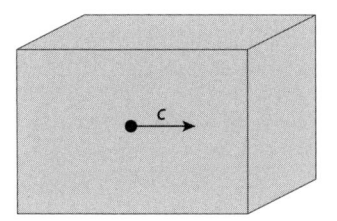

When the molecule collides with the right-hand wall it will rebound with velocity $-c$.

change in momentum = $-2mc$

The molecule travels a distance of $2L$ before colliding with that wall again, so the time elapsed is $2L/c$.

rate of change of momentum = force applied by the molecule on this wall = $\dfrac{2mc}{2L/c} = \dfrac{mc^2}{L}$

The area of the wall is L^2, so

pressure = force/area = $\dfrac{mc^2}{L^3}$

The molecule being considered is moving perpendicular to the two faces with which it collides. In practice, a typical molecule moves randomly and collides with all six faces. Thus, the total area involved is three times that which has been considered, so

pressure = $\dfrac{mc^2}{3L^3}$

The total number of molecules in the box is N, each with a different speed c contributing to the overall pressure. The average of the velocities squared is called the mean square velocity, $<c^2>$.

So $p = \dfrac{1}{3}\dfrac{Nm<c^2>}{L^3}$, $L^3 = V$, the volume of the box.

$$p = \dfrac{1}{3}\dfrac{Nm<c^2>}{V}.$$

It is sometimes useful to write this equation as:

$pV = \tfrac{1}{3}Nm<c^2>$

Worked example

At room temperature and pressure (293 K and 1.0×10^5 Pa), 1 mol of any gas occupies a volume of 24 dm³.

Calculate the root mean square velocity of the following at this temperature:

(a) helium atoms (atomic mass = 4 u)

(b) oxygen molecules (atomic mass = 16 u, mass of O_2 = 32 u)

Answer

(a) Nm = total mass of 1 mol of helium = 4×10^{-3} kg

$$p = \frac{1}{3}\frac{Nm<c^2>}{V} \text{ therefore}$$

$$<c^2> = \frac{3pV}{Nm} = \frac{3 \times 1.0 \times 10^5 \times 24 \times 10^{-3}}{4 \times 10^{-3}} = 1.8 \times 10^6 \, m^2 s^{-2}$$

$\sqrt{<c^2>} = 1342 \, m\,s^{-1} = 1300 \, m\,s^{-1}$ (2 s.f.)

(b) Nm = total mass of 1 mol of O_2 molecules = 3.2×10^{-2} kg

$$<c^2> = \frac{3pV}{Nm} = \frac{3 \times 1.0 \times 10^5 \times 24 \times 10^{-3}}{3.2 \times 10^{-2}} = 2.25 \times 10^5 \, m^2 s^{-2}$$

$\sqrt{<c^2>} = 470 \, m\,s^{-1}$

Note

The root mean square velocity is the square root of the mean square velocity.

Temperature and molecular kinetic energy

If we compare the ideal gas equation ($pV = nRT$) and the equation $pV = \frac{1}{3}Nm<c^2>$ we can see that:

$nRT = \frac{1}{3}Nm<c^2>$

For one mole:

$$\frac{R}{N_A}T = \frac{1}{3}m<c^2>$$

where R/N_A is known as the Boltzmann constant (k), which has the value $1.38 \times 10^{-23} \, J\,K^{-1}$.

$kT = \frac{1}{3}m<c^2>$ and hence

$$\tfrac{3}{2}kT = \tfrac{1}{2}m<c^2>$$

$\frac{1}{2}m<c^2>$ is equal to the average (translational) kinetic energy of a molecule. Hence the temperature is proportional to the average (translational) kinetic energy of the particles in a gas.

Temperature

What is temperature?

You have been using the idea of temperature for many years and will have an instinctive feeling about its meaning. However, that instinctive feeling may not be fully correct. We saw in the last section that temperature is proportional to the average kinetic energy of the molecules in a body. To take this a stage further, temperature tells us the direction in which there will be a net energy flow between bodies in thermal contact: energy will tend to flow from a body at high temperature to a body at a lower temperature. If there is no net energy flow between two bodies in thermal contact then those two bodies will be at the same temperature. They are said to be in thermal equilibrium.

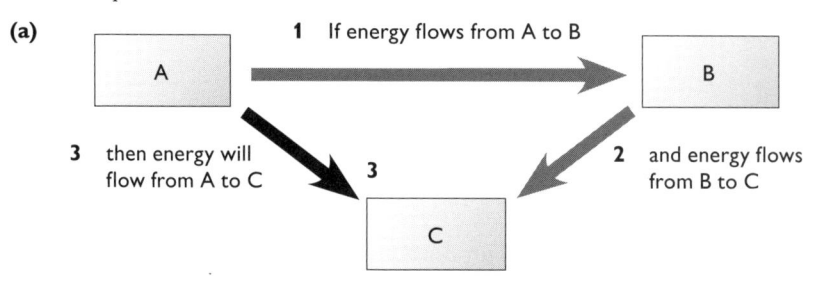

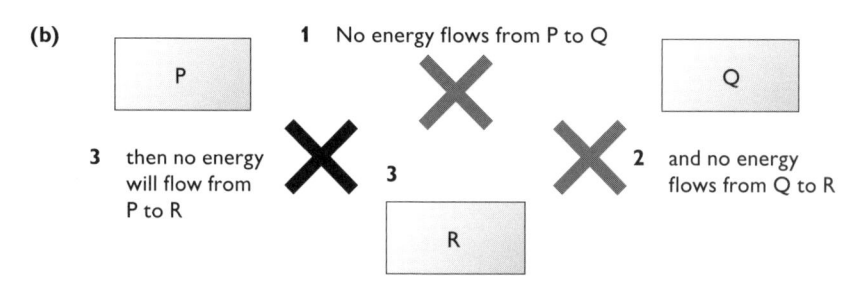

The energy flow between different bodies in thermal contact

Diagram (a) above shows that if body A is at a higher temperature than B and if body B is at a higher temperature than body C, then body A is at a higher temperature than body C.

Diagram (b) shows that if body P is in thermal equilibrium with body Q and if body Q is in thermal equilibrium with body R, then body P is in thermal equilibrium with body R.

Measurement of temperature

To measure temperature, a physical property that varies with temperature is used. Examples are:
- expansion of a liquid
- expansion of a gas at constant pressure
- change of pressure of a gas at constant volume
- change in resistance of a metallic conductor
- change in resistance of a thermistor or other semiconductor
- e.m.f. produced across the junctions of a thermocouple

Electrical thermometers
The thermocouple

Copper wire — Iron wire — Junction 1 — Junction 2 — Copper wire — V

The diagram shows the structure of a thermocouple. The wires need not be copper and iron — any two different metals can be used. When the two junctions are at different temperatures an e.m.f. is produced, which is measured by the voltmeter. The e.m.f. increases with increasing temperature difference. It may not change linearly with temperature, in which case a calibration graph will have to be used (see pages 23–24).

In practice, one junction is kept at a constant temperature, perhaps in melting ice, while the other acts as the 'test junction'.

Resistance thermometers

There are two types of resistance thermometer — metallic type and semiconductor type. The latter is usually referred to as a thermistor. Both types rely on the change of resistance as the temperature changes. The resistance of a metal increases with increasing temperature; the resistance of a semiconductor (thermistor) generally decreases with increasing temperature (see page 85). For both types of thermometer a calibration curve must be drawn.

Thermocouples and resistance thermometers are compared in Table 15.

Table 15

Property	Metallic resistance	Thermistor	Thermocouple
Range	Wide	Narrow	Wide
Reaction rate	Slower than the thermocouple, larger in size and therefore larger thermal capacity	Slower than the thermocouple, larger in size and therefore larger thermal capacity	Fast, the active part of the thermometer is the 'test junction', which is small and therefore has a low thermal capacity. Can measure the temperature of very small objects and at a 'point' on a larger object
Sensitivity	Less sensitive than the thermistor	Sensitive over a narrow range	Can be extremely sensitive, depending on choice of metals
Remoteness	The wires to the thermometer can be long, therefore the operator can be remote	The wires to the thermometer can be long, therefore the operator can be remote	The wires to the thermometer can be long, therefore the operator can be remote

The thermodynamic, or Kelvin, temperature scale

All temperature scales require two fixed points that are easily repeatable. For example, the Celsius scale uses the melting point of pure water as the lower fixed point (0 °C) and the boiling point of pure water at standard atmospheric pressure as the higher fixed point (100 °C). When a thermometer is calibrated, these two points are marked and then the scale is divided into 100 equal parts.

The thermodynamic temperature scale is independent of the physical properties of any particular substance. The fixed points it takes are:
- **absolute zero** — this is the temperature at which no more energy can be removed from any body, all the energy that can be removed has been removed. At this temperature all substances have minimum internal energy. This does *not* mean that the body has zero energy; the random kinetic energy of the molecules is zero and the potential energy is a minimum. Only an ideal gas will have zero energy; it will also have zero pressure. (0 K = −273.15 °C)
- the **triple point of pure water** — the unique temperature at which water exists in equilibrium as a vapour, a liquid and a solid. (273.16 K = 0.01 °C)

For convenience, the size of the unit in the thermodynamic scale was chosen to be the same size as the degree in the Celsius scale. You will see the triple point is just above the melting point of water (0 °C).

Conversion between the Celsius and the Kelvin scales

$T/K = T/°C + 273.15$

In practice, we often simplify the conversion by using $T K = T °C + 273$

Worked example
Copy and complete the table, showing your working.

	Temperature/K	Temperature/°C
Boiling point of water		100
Boiling point of bromine	332.40	
Boiling point of helium	4.37	
Melting point of hydrogen		−258.98
Boiling point of nitrogen	77.50	

Answer

	Temperature/K	Temperature/°C
Boiling point of water	100 + 273 = 373	100
Boiling point of bromine	332.40	332.40 − 273.15 = +59.25
Boiling point of helium	4.37	4.37 − 273.15 = −268.78
Triple point of hydrogen	259.34 − 273.15 = 13.81	−259.34
Boiling point of nitrogen	77.50	77.50 − 273.15 = −195.65

III Matter

> **Note**
> The boiling point of water is given only to the nearest degree Celsius. Therefore, using 273 as the difference between Celsius and Kelvin is justified.

Thermal properties of materials

Specific heat capacity

When a body is heated its temperature increases. The amount that it increases by (ΔT) depends on:
- the energy supplied (ΔE)
- mass of the body (m)
- the material the body is made from

$\Delta T \propto \frac{\Delta E}{m}$, which can be written $\Delta E = mc\Delta T$

c is the constant of proportionality. Its value depends on the material being heated. It is known as the specific heat capacity of the material.

Rearranging the equation gives: $c = \frac{\Delta E}{m \Delta T}$

The specific heat capacity of a material is defined as the energy required to raise the temperature of unit mass of the material by 1 K (or 1 °C).

The units of specific heat capacity are $J\,kg^{-1}\,K^{-1}$.

This is often written as $J\,kg^{-1}\,°C^{-1}$. The units are numerically equal.

Worked example

A block of aluminium has a mass of 0.50 kg. It is heated, with a 36 W heater, for 3 minutes and its temperature increases from 12°C to 26°C.

Calculate the specific heat capacity of aluminium.

Answer

$c = \frac{\Delta E}{m \Delta T} = \frac{36 \times 3 \times 60}{0.50 \times 14} = 930\,J\,kg^{-1}\,°C^{-1}$

This is slightly higher than the recognised figure. However, there is no attempt to allow for energy losses to the surroundings.

Specific heat capacity and the kinetic theory

In the earlier section on the kinetic theory, we met the idea that the temperature of a gas is a measure of the kinetic energy of the molecules. This theory can be extended to both liquids and solids — when the temperature of any body is increased the average kinetic energy of the particles in the body is increased.

Measurement of specific heat capacity

The principles of measuring the specific heat capacity of either a solid or liquid are simple:
- find the mass of material being heated
- measure the energy input
- measure the temperature change

The apparatus required for a straightforward experiment is shown in the diagram below.

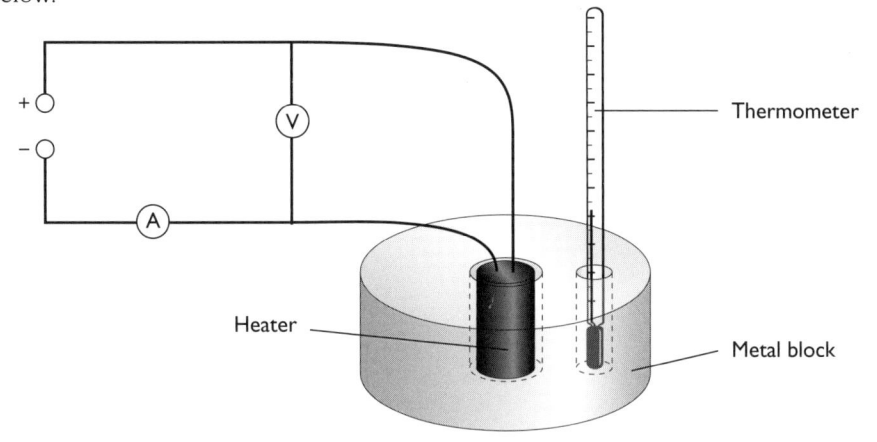

This is the sort of experiment that you may have met in earlier years of study. The mass and temperature change of the block are measured using a balance and thermometer. The energy input can be calculated form the power input (VI) multiplied by the time for which the heater is switched on.

The major problem with this experiment is to measure and/or reduce energy losses to the surroundings. Simple precautions can be taken:
- Insulate the block.
- Start the experiment with the block below room temperature and then turn the heater off when the block is at an equivalent temperature above room temperature. The block will gain energy from the surroundings when it is below room temperature and lose an equal amount when it is above room temperature.

When similar experiments are carried out to find the specific heat capacity of liquids, it must be remembered that the container that holds the liquid also requires energy to raise its temperature. Often, containers made from expanded polystyrene or similar insulating materials are used. These have two advantages:
- They provide the necessary insulation to reduce energy losses.
- They have a low **thermal capacity**.

The thermal capacity of a body is the energy required to raise the temperature of the complete body by 1 °C.

thermal capacity, $C = mc$ where m = mass of the body and c = the specific heat capacity of the body.

Worked example
An electric shower is designed to work from a 230 V mains supply. It heats water as it passes through narrow tubes prior to the water passing through the shower head. Water enters the heater at 12 °C and when the flow rate is 0.12 kg s^{-1} it leaves at 28 °C.

Calculate the current in the heater, assuming that energy losses are negligible.

(specific heat capacity of water = 4200 J kg^{-1} °C^{-1})

Answer
power = $VI = mc\Delta T$, where m is the mass of water passing through the heater per second

$230I = 0.12 \times 4200 \times (28 - 12)$

$I = 35$ A

Note

It is worth noting that this is a fairly high current and that electric showers tend to be on a separate circuit from other appliances. It is also worth recording that one way of adjusting the temperature of the shower is to alter the flow rate.

Latent heat

You will have observed that when ice melts (or water boils) the ice (or boiling water) remains at a constant temperature during the process of melting (or boiling), despite energy still being supplied. This energy does not change the temperature of the substance. Instead, it is doing work in changing the solid to liquid (or liquid to vapour). This energy is called the **latent heat** of fusion (or vaporisation).

The **specific latent heat of fusion** is the energy required to change unit mass of solid to liquid without change in temperature.

The **specific latent heat of vaporisation** is the energy required to change unit mass of liquid to vapour without change in temperature.

From the definitions:

$L_f = \dfrac{\Delta E}{\Delta m}$, where L_f is the specific latent heat of fusion, ΔE is the energy input, and Δm is the mass of solid converted to liquid.

$L_v = \dfrac{\Delta E}{\Delta m}$, where L_v is the specific latent heat of vaporisation, ΔE is the energy input, and Δm is the mass of liquid converted to vapour.

The units of both specific latent heat of fusion and of vaporisation are J kg^{-1}.

Worked example
A 1.5 kW kettle contains 400 g of boiling water.

Calculate the mass of water remaining if it is left switched on for a further 5 minutes.

(specific latent heat of vaporisation of water = 2.26 MJ kg^{-1})

Answer
$$L_v = \frac{\Delta E}{\Delta m}$$
Therefore $\Delta m = \frac{\Delta E}{L_v} = \frac{1.5 \times 10^3 \times 5 \times 60}{2.26 \times 10^6} = 0.199 \text{ kg} = 199 \text{ g}$

mass remaining = 400 − 199 = 201 g

Latent heat and kinetic theory
We have seen how the average kinetic energy of the particles of a body increases as the temperature increases. When there is a change of state there is no change in the kinetic energy of the particles — this is why there is no change in the temperature. Instead, work is done against the interparticle forces in separating the particles.

The average increase in the separation of particles when a solid turns to a liquid is small, although the interparticle forces are relatively large. The particles will now have more potential energy than in the solid state.

The specific latent heat of vaporisation of a substance is generally greater than the specific latent heat of fusion because in melting, work is only done against two or three bonds whereas in vaporisation work is done against up to a dozen bonds.

Measurement of the specific latent heat of fusion of ice
The diagram below shows apparatus which could be used to measure the specific latent heat of fusion of ice.

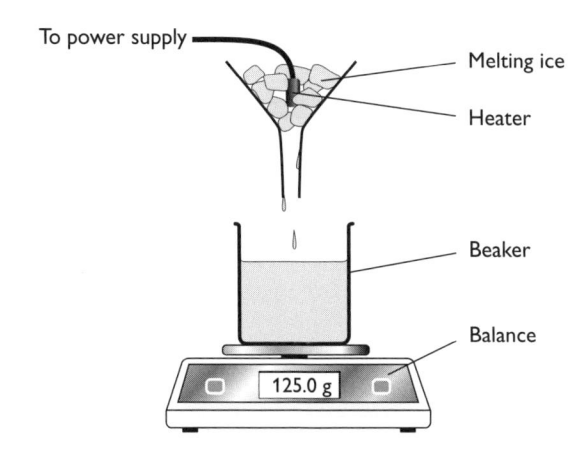

Determining the specific latent heat of fusion of ice

III Matter

The method is straightforward. The heater melts the ice and the resulting water is collected in the beaker.

If the power of the heater is P, the mass of the beaker before the heater is switched on is m_1, the mass of the beaker plus water is m_2, and the heater is switched on for time t, then:

the latent heat of fusion of water, $L_f = \dfrac{P \times t}{m_2 - m_1}$

This, however, does not take into account energy exchanges with the surroundings. In this case, because the melting point of ice is less than room temperature it will be an energy gain, rather than a loss. One method to allow for this is to find the mass of the water collected for a given time before the heater is switched on. This gives the mass of water melted by the energy received from the surroundings. This can be subtracted from the mass collected when the heater is switched on. The following example demonstrates this.

Worked example

An experiment is carried out to measure the specific latent heat of fusion of water using the apparatus shown on page 155. The power of the heater is 48 W. The results obtained are shown in the table.

	Initial reading on the balance/g	Final reading on the balance/g	Time/s
Heater off	116.2	124.4	480
Heater on	124.4	164.4	240

Calculate the specific latent heat of fusion of water.

Answer

ice melted due to energy gained from the surroundings = 124.4 − 116.2 = 8.2 g

ice melted due to this energy gained during the experiment = 8.2/2 = 4.1 g

ice melted during the heating = 164.4 − 124.4 = 40.0 g

ice melted due to the heater = 40.0 − 4.1 = 35.9 g

$L_f = \dfrac{P \times t}{m_2 - m_1} = \dfrac{48 \times 240}{35.9} = 320 \, \text{J g}^{-1} = 3.2 \times 10^5 \, \text{J kg}^{-1}$

Measurement of the specific latent heat of vaporisation of water

A similar method can be used to measure the specific latent heat of vaporisation of water.

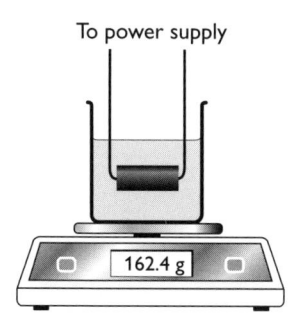

Determining the specific latent heat of vaporisation of water

The problem, in this case, is to allow for energy losses to the surroundings. This can be done by repeating the experiment using a heater of different power for the same length of time as in the original experiment. The energy losses will be equal each time, so the differences between the two sets of results will cancel out the effect of heat losses. This is shown in the following worked example.

Worked example
An experiment is carried out to measure the specific latent heat of vaporisation of water using the apparatus shown in the diagram above.

The results obtained are shown in the table.

	Power of heater/W	Time/s	Mass of beaker and water at the beginning of the experiment/g	Mass of beaker and water at the end of the experiment/g
Experiment 1	36	600	177.8	168.4
Experiment 2	50	600	168.4	155.3

Calculate the specific latent heat of vaporisation of water.

Answer
Experiment 1:
input energy $- q = \Delta m L_v$ where q = energy lost to the surroundings
$36 \times 600 - q = (177.8 - 168.4)L_v$
$21\,600 - q = 9.4 L_v$

Experiment 2:
input energy $- q = \Delta m L_v$ where q = energy lost to the surroundings
$50 \times 600 - q = (168.4 - 155.3)L_v$
$30\,000 - q = 13.1 L_v$

Subtract the first equation from the second:
$30\,000 - 21\,600 = (13.1 - 9.4)L_v$
$8400 = 3.7 L_v$
$L_v = 2.27 \times 10^3 \, \text{J g}^{-1}$

Evaporation and boiling

You should be aware of the difference between evaporation and boiling.

Table 16

Evaporation	Boiling
Molecules escape from the surface of the liquid	Bubbles of vapour form in the body of the liquid
Takes place over a wide range of temperatures	Takes place at a single temperature

It is often incorrectly thought that evaporation does not require energy input. The confusion may arise because water evaporates at room temperature. However, we can see that the cooling that evaporation causes is likely to reduce the temperature of the water to below room temperature. Consequently there will be an energy transfer from the surroundings to the water (see page 149).

Evaporation causes cooling. This can be understood if we consider kinetic theory. This is illustrated in the diagram below.

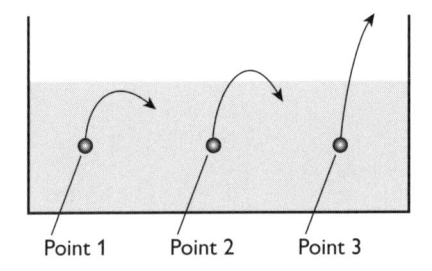

Point 1   Point 2   Point 3

In the diagram:
- At point 1— a slow-moving molecule approaches the surface of a liquid but is pulled back into the body of the liquid.
- At point 2 — a slightly faster-moving molecule approaches the surface of a liquid, just gets out of the surface, but there is still sufficient attraction to pull the molecule back into the liquid.
- At point 3 — a fast-moving molecule approaches the surface of a liquid. There is a tendency for it to be pulled back into the liquid, but it has sufficient energy to escape.

Molecules escape from the surface of the liquid. When a molecule is in the body of the liquid, the net force on it is zero because the pull from all the molecules around it cancel each other out. However, when a molecule approaches the surface of a liquid there is a resultant force towards the centre of the liquid because there are very few molecules above it. Consequently, only the fastest-moving molecules can escape from the surface. This means that the average speed of those left behind falls. Remember, the temperature is a measure of the average speed of the molecules in a body. When the average speed drops, the temperature drops.

Internal energy

In the previous sections we have seen that the particles in a body have a mixture of kinetic energy and potential energy. Kinetic energy determines the temperature of the body and potential energy determines the state of the body. Not all particles have the same kinetic and potential energies, they are randomly distributed. The internal energy of a body is the sum of the kinetic and potential energies of all the particles in the body.

Internal energy is the sum of the random distribution of the kinetic and potential energies associated with the particles of a system.

There are two ways of increasing the total internal energy of a body:
- heating the body
- doing work on the body

This leads to the first law of thermodynamics, which can be expressed by the equation:

increase in internal energy (ΔU) = the energy supplied to the system by heating (Q) + the work done on the system (W)

$\Delta U = Q + W$

(The 'energy supplied to the system by heating' is sometimes shortened to the 'heating of the system'.)

To demonstrate a use of the first law, consider an ideal gas contained in a cylinder by a frictionless piston. The initial volume of gas is V.

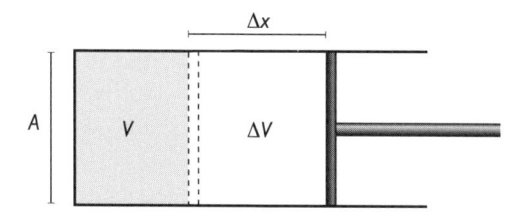

The gas is heated so that its volume increases by an amount ΔV against a constant atmospheric pressure.

The gas expands, so work is done against atmospheric pressure:

$W = F\Delta x$, where F is the force on the piston

force on the piston = pressure of the gas × area of cross-section of the piston = pA

so, $W = pA\Delta x = p\Delta V$

Applying the first law of thermodynamics:

$\Delta U = Q + W$

$\Delta U = Q - p\Delta V$

The minus sign comes in because the work is done by the gas on the atmosphere, rather than work being done on the gas. You will notice that the change in internal energy is less than the energy input ΔQ, because some of the energy is used to do work in expanding the gas.

Table 17 explains when the quantities, ΔU, Q and W should be considered positive and negative.

Table 17

Quantity	Positive	Negative
ΔU	The internal energy increases	The internal energy decreases
Q	Energy is transferred to the system from the surroundings by heating	Energy is transferred from the system to the surroundings by heating
W	Work is done on the system	Work is done by the system

Adiabatic expansion and compression

This is an expansion or compression when no energy leaves or enters a gas. In this case $Q = 0$, so

$\Delta U = W$

This means that when a gas expands and it does work on the atmosphere, W is negative. Therefore, ΔU is negative and the gas cools down. You may have observed that when carbon dioxide is released from a high-pressure cylinder, it cools so much that solid carbon dioxide (dry ice) is formed. If the gas is compressed, work is done on the gas and the gas is warmed. This is observed when pumping up a bicycle tyre — the barrel of the pump gets much hotter than it would from just doing work against friction.

Oscillations and waves

Oscillations

Terminology

Consider a ruler clamped to a bench, pulled downwards and released so that it vibrates, a pendulum swinging backwards and forwards or a mass on the end of a spring bouncing up and down. These are all examples of oscillating systems.

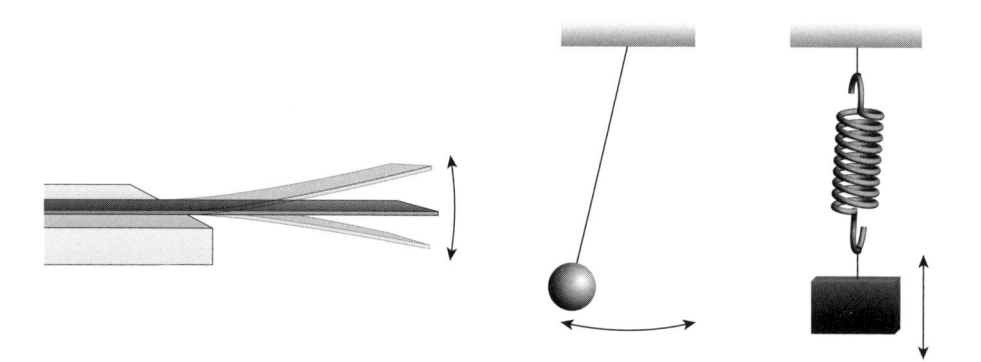

One complete **oscillation** is when a particle moves from its equilibrium position, to its maximum displacement in one direction, back through the equilibrium position to the maximum displacement in the opposite direction and back once more to the equilibrium position. This is shown in the diagram below.

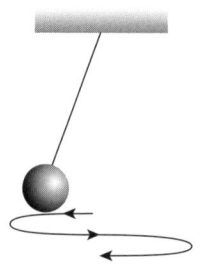

One complete oscillation

- The **period**, *T*, is the time taken for one complete oscillation of a particle.
- The **frequency**, *f*, is the number of oscillations per unit time.
- The **amplitude**, x_0, is equal to the magnitude of the maximum displacement of a particle from its mean position.
- The **angular frequency**, Ω or ω, is equal to $2\pi f$.

It is worth remembering the following relationships, which you may recognise from the work on circular motion:

$f = 1/T$

$\omega = 2\pi f$

$\omega = 2\pi/T$

Simple harmonic oscillations

In the examples above, the bodies vibrate in a particular way known as **simple harmonic motion (shm)**. There are many other types of oscillation. For instance, a conducting sphere will oscillate between two charged conducting plates — however this is not a simple harmonic oscillation.

IV Oscillations and waves

The conditions required for simple harmonic motion are:
- vibration of a particle about a fixed point
- the acceleration is always directed towards that fixed point
- the magnitude of the acceleration is proportional to the displacement from that fixed point

Simple harmonic motion can be investigated using a position sensor connected to a datalogger.

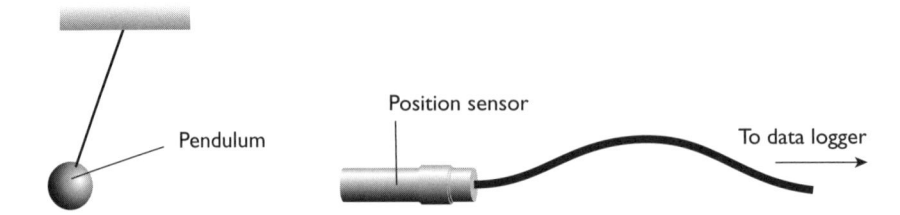

The displacement against time graph can be deduced from the trace from the datalogger:

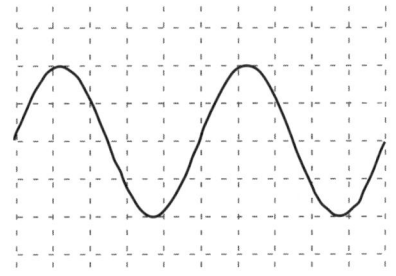

As with any displacement–time graph the velocity is equal to the gradient of the graph; the acceleration is equal to the gradient of the velocity–time graph.

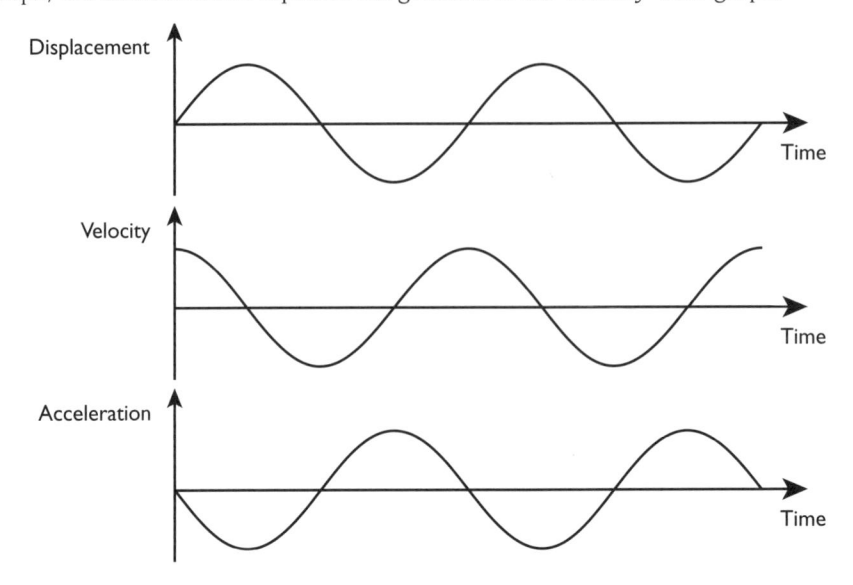

Table 18 describes the displacement, the velocity and the acceleration at different points during an oscillation with reference to these graphs.

Table 18

Point in cycle	Displacement	Velocity	Acceleration
t = 0	Zero	Maximum in one direction	Zero
¼ cycle on from t = 0	Maximum in one direction	Zero	Maximum in the opposite direction to the displacement
½ cycle on from t = 0	Zero	Maximum in the opposite direction to before	Zero
¾ cycle on from t = 0	Maximum in the opposite direction to before	Zero	Maximum in the opposite direction to the displacement
1 cycle on from t = 0	Zero	Maximum in the original direction	Zero

You can now see why the oscillation of a conducting sphere bouncing between two charged plates is not simple harmonic.

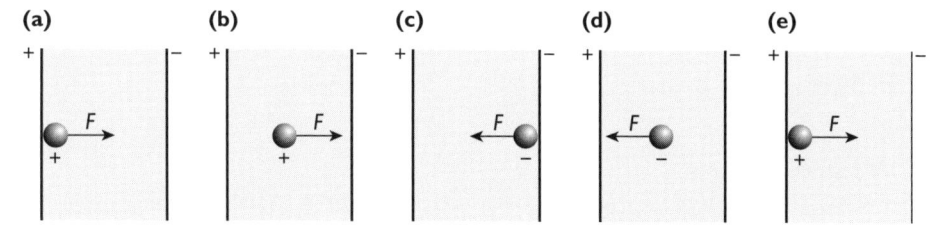

A sphere bouncing between two charged plates

(a) The sphere touches the positive plate and becomes positively charged — there is a force away from the positive plate towards the negative plate
(b) The force on the charge is constant all the way between the plates
(c) The sphere touches the negative plate and becomes negatively charged — there is a force away from the negative plate towards the positive plate
(d) The force on the charge is constant all the way between the plates
(e) Completes the oscillation, the sphere becomes positively charged once more

It is clear that the force on the sphere and, hence, the acceleration is not directed towards a fixed point. The acceleration is constant as the sphere travels between the plates, then suddenly changes to the opposite direction when it hits a plate.

Equations for simple harmonic motion

If you look at the graphs of simple harmonic motion you will see that they are of the form of sine (or cosine) graphs. The conditions for shm give the following proportionality:

$a \propto -x$ where a is the acceleration and x is the displacement

163

IV Oscillations and waves

The minus sign comes in because the acceleration is in the opposite direction to the displacement.

This leads to the equation:

$a = -\omega^2 x$, where ω is the angular frequency

This equation describes simple harmonic motion. The graphs on page 162 are 'solutions' to this equation. If you look at those graphs you will see that they have a sine (or cosine) shape. The precise equations that they represent are:
- displacement: $x = x_0 \sin \omega t$
- velocity: $v = x_0 \omega \cos \omega t$
- acceleration: $a = -x_0 \omega^2 \sin \omega t$

x_0 is the amplitude of the oscillation.

If you look at the equations for displacement and acceleration you should be able to see that they fit in with the equation $a = -\omega^2 x$.

The velocity of the vibrating body at any point in the oscillation can be calculated using the formula:

$v = \pm \omega \sqrt{x_0^2 - x^2}$

It follows that when $x = 0$ (i.e. the displacement is zero) the velocity is a maximum and

$v = \pm \omega x_0$

Worked example
A mass on the end of a spring oscillates with a period of 1.6 s and an amplitude of 2.4 cm.

Calculate:
- (a) the angular frequency of the oscillation
- (b) the maximum speed of the mass
- (c) the maximum acceleration
- (d) the speed of the particle when its displacement from the equilibrium position is 0.6 cm

Answer
(a) $f = 1/T = 1/1.6$ Hz
   $\omega = 2\pi f = 2\pi \times (1/1.6) = 3.9$ rad s^{-1}

(b) $v_{max} = \omega x_0 = 3.9 \times 2.4 = 9.4$ cm s^{-1}

(c) $a = -\omega^2 x$
   Therefore, $a_{max} = \omega^2 x_0 = 3.9^2 \times 2.4 = 37$ cm s^{-2}

(d) $v = \omega \sqrt{x_0^2 - x^2} = 3.9 \times \sqrt{2.4^2 - 0.6^2}$
   $v = 9.1$ cm s^{-1}

International AS and A Level Physics Revision Guide

> **Note**
>
> It is worth noting that the displacement–time graph can be started at any point on the cycle. This text has chosen the equilibrium position as the starting point. Other texts might choose maximum displacement, in which case the displacement curve would be a cosine curve, the velocity curve would be a minus sine curve and the acceleration would be a minus cosine curve.

shm and circular motion

The introduction of ω should have reminded you of circular motion. The experiment described below shows the relationship between circular motion and simple harmonic motion.

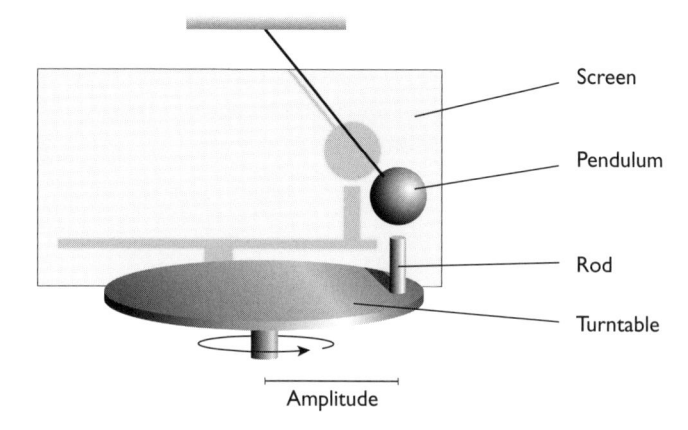

A rod is set up on a turntable which rotates. A pendulum is set swinging with an amplitude equal to the radius of the rotation of the rod. The speed of rotation of the turntable is adjusted until the time for one revolution of the turntable is exactly equal to the period of the pendulum. The whole apparatus is illuminated from the front so that a shadow image is formed on a screen.

It is observed that the shadow of the pendulum bob moves exactly as the shadow of the rod. This shows that the swinging of the pendulum is the same as the projection of the rod on the diameter of the circle about which it rotates.

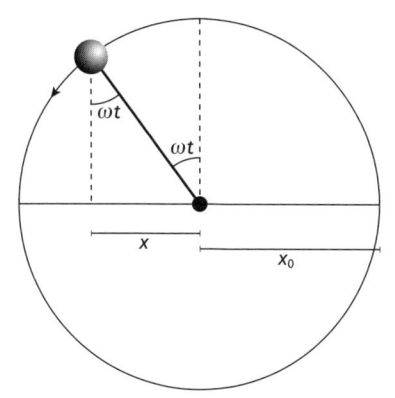

You should now understand the close relationship between circular motion and shm.

Energy and shm

During simple harmonic motion, energy is transferred continuously between kinetic and potential energy:
- In the case of a pendulum, the transfer is between kinetic and gravitational potential energy.
- In the case of a mass tethered between two horizontal springs, the transfer is between kinetic to strain potential energy.

The important point is that in any perfect simple harmonic oscillator the total energy is constant. This means that the sum of the kinetic and potential energies remains constant throughout each oscillation.

The speed of the particle is at a maximum when the displacement is zero so that the kinetic energy is maximum at this point and the potential energy is zero. At maximum displacement the speed, and hence the kinetic energy, is zero and the potential energy is maximum.

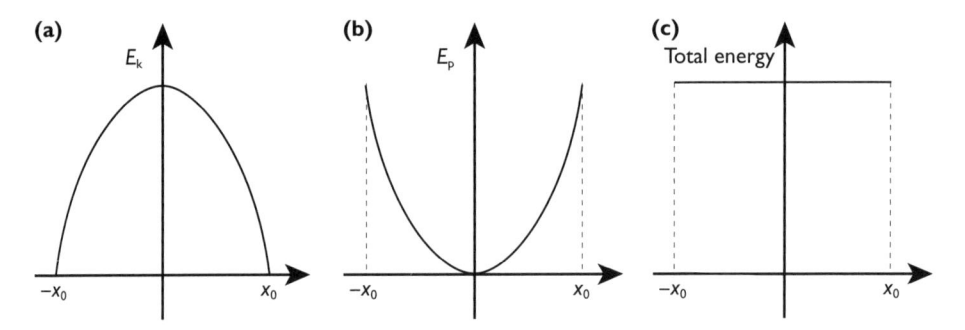

(a) The variation of kinetic energy with displacement
(b) The variation of potential energy with displacement
(c) The total energy with displacement

The equations that link the kinetic energy and the potential energy to the displacement are:
- kinetic energy: $E_k = \frac{1}{2}m\omega^2(x_0^2 - x^2)$
- potential energy: $E_p = \frac{1}{2}m\omega^2 x^2$

Worked example

A clock pendulum has a period of 2.0 s and a mass of 600 g. The amplitude of the oscillation is 5.2 cm.

Calculate the maximum kinetic energy of the pendulum and, hence, its speed when it is travelling through the centre point.

Answer

$E_k = \frac{1}{2}m\omega^2(x_0^2 - x^2)$ for maximum speed the displacement = 0

$T = 2.0$, therefore $\omega = 2\pi/2 = \pi$

$E_k = \frac{1}{2}m\omega^2 x_0^2 = 0.5 \times 0.60 \times \pi^2 \times (5.2 \times 10^{-2})^2$

$E_k = 8.0 \times 10^{-3}\,\text{J}$

$E_k = \frac{1}{2}mv^2$

$v = \sqrt{2E_k/m} = \sqrt{2 \times 8.0 \times 10^{-3}/0.60} = \sqrt{0.027}$

$= 0.16\,\text{m s}^{-1}$

Note

If you use the formula $v = \omega x_0$, you will find that the answer comes to $16\,\text{cm s}^{-1}$, which is in agreement with this value.

Damping

Up to this point we have only looked at perfect simple harmonic motion, where the total energy is constant and no energy is lost to the surroundings. In this situation, where the only force acting on the oscillator is the restoring force, the system is said to be in **free oscillation**. In real systems, some energy is lost to the surroundings due to friction and/or air resistance. This always acts in the opposite direction to the restoring force. The result is that the amplitude of the oscillations gradually decreases. This is called **damping**.

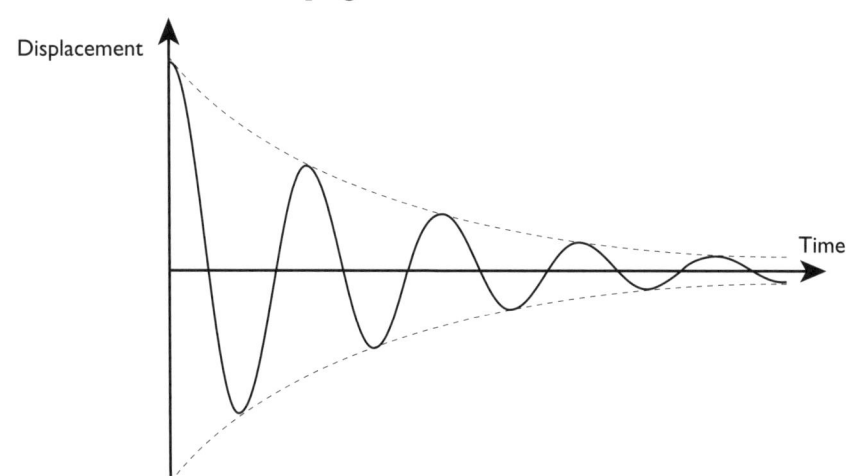

The decay of the oscillation follows the exponential decay (see page 219). The period, however, remains constant until the oscillation dies away completely. The diagram above shows **light damping** — the oscillation gradually fades away. If the damping is increased we eventually reach a situation where no complete oscillations occur and the displacement falls to zero. When this occurs in the minimum time, the damping is said to be **critical.** More damping than this is described as **overdamping** and the displacement only slowly returns to zero.

IV Oscillations and waves

Examples of damped oscillations

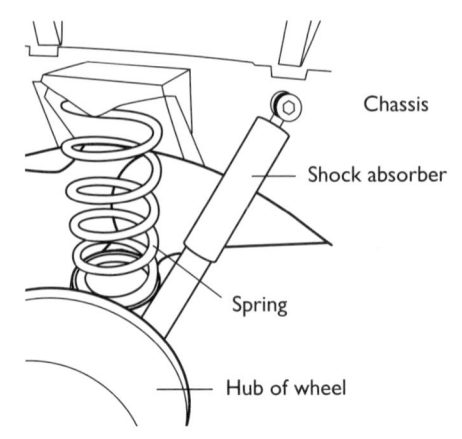

Chassis
Shock absorber
Spring
Hub of wheel

A car suspension operates in a critical damping mode in order to bring the displacement back to zero in the shortest possible time without oscillations. An overdamped suspension leads to a hard ride with energy given to the car by bumps not being absorbed as efficiently.

Forced oscillations

At AS you met the idea of stationary waves formed on a string (page 72). This is an example of a **forced oscillation**; an extra periodic force is applied to the system. This periodic force continuously feeds energy into the system to keep the vibration going.

You will have observed how the amplitude of the vibrations of the waves on a string changes as the frequency of the vibrator is increased — small amplitude at very low frequencies gradually increasing to a maximum as the frequency is increased, then reducing again as the frequency is increased further.

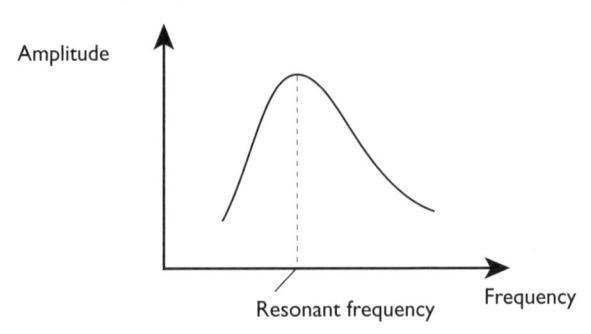

This is an example of **resonance**. If the driving frequency is the same as the natural frequency of oscillation of the string, then it gives the string a little kick at the right time each cycle and the amplitude builds up.

Resonance can be demonstrated using Barton's pendulums.

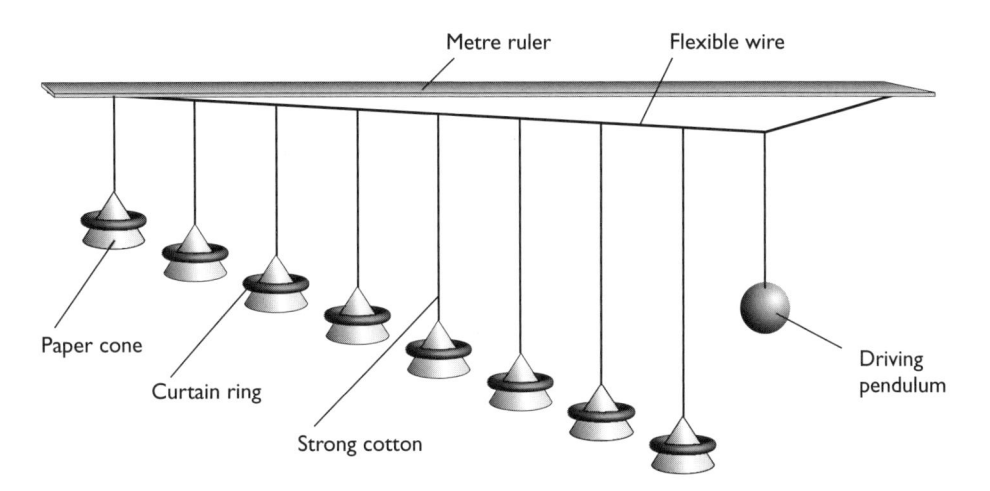

The driving pendulum causes the paper-cone pendulums to vibrate. Only the pendulums of a similar length to the driving pendulum show any significant oscillation. All the pendulums vibrate with the same frequency, which is the frequency of the driving pendulum (not their own natural frequencies). This is a general rule for all forced oscillations.

Resonance and damping

Resonance can be useful. For instance, whenever a trombone or other wind instrument is played, stationary waves — which are an example of resonance — are set up. Whenever you listen to a radio you are relying on resonance because the tuning circuit in the radio will have the same natural frequency as the radio wave that transmits the signal.

However, resonance can be a problem in machinery, in bridge and building design, and in aircraft. Large oscillations of aircraft parts, for example, can cause 'metal fatigue', which leads to early failure of those parts.

In order to reduce resonance peaks, damping is introduced. This has the desired effect of reducing the amplitude of vibrations. However, it also tends to spread the range of frequencies at which large vibrations occur.

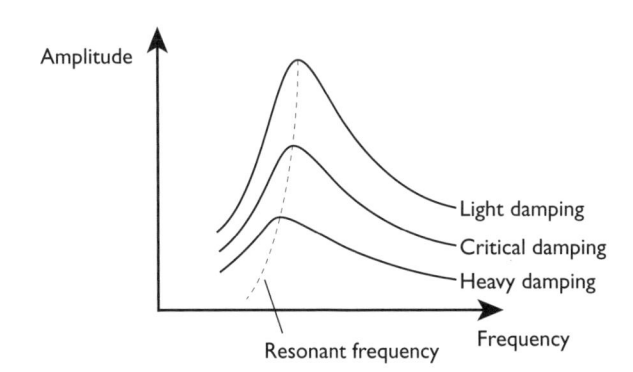

This can be seen by removing the curtain rings from the paper cones in Barton's experiment. This reduces the mass of the pendulums, meaning that air resistance has a greater damping effect. It can be seen that there is a greater amplitude of vibration of those pendulums near but not at the resonant frequency.

There is a further effect as the damping of an oscillator is increased — the frequency at which the maximum amplitude occurs (the resonant frequency) falls slightly.

Electricity and magnetism
Electric fields

We have already looked at uniform electric fields in the AS course (pages 80–82). It would be a good idea to revise those ideas before continuing with this section. The ideas introduced earlier are developed and non-uniform fields are introduced.

Coulomb's law

We have already met the idea that unlike charges attract and like charges repel. You may have deduced that the sizes of the forces between charges depend on:
- the magnitude of the charges on the two bodies; $F \propto Q_1 Q_2$, where Q_1 and Q_2 are the charges on the two bodies
- the distance between the two bodies; $F \propto 1/r^2$ where r is the distance between the (centres of) the two charged bodies

This leads to $F \propto \dfrac{Q_1 Q_2}{r^2}$.

The constant of proportionality is $1/4\pi\varepsilon_0$, where ε_0 is known as the permittivity of free space and has a value $8.85 \times 10^{-12}\,\text{C}^2\,\text{N}^{-1}\,\text{m}^{-2}$. This unit is often shortened to farads per metre ($\text{F}\,\text{m}^{-1}$). We will meet the farad in the section on capacitors (see page 175):

$$F = \dfrac{Q_1 Q_2}{4\pi\varepsilon_0 r^2}$$

This relationship only really applies when point charges are considered. However, the charge on a spherical body can be considered to act at the centre of the body, provided that the distance between the two bodies is considerably greater than their diameters.

> **Worked example**
> The diagram shows how the forces on two charged bodies can be investigated. The two conducting spheres are identical, each having a diameter of 1.2 cm. They are charged by connecting to the same very high voltage supply.

Use the data in the diagrams to find the charges on the two spheres.

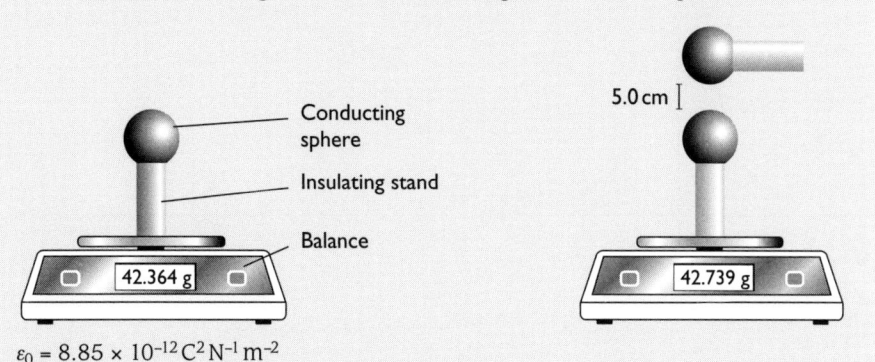

[Diagram: Two conducting spheres on insulating stands on balances. Left balance reads 42.364 g. Right balance reads 42.739 g. Spheres separated by 5.0 cm.]

$\varepsilon_0 = 8.85 \times 10^{-12}\,\text{C}^2\,\text{N}^{-1}\,\text{m}^{-2}$

Answer

distance between centres of the spheres = $5.0 + (½ \times 2 \times 1.2) = 6.2 \times 10^{-2}\,\text{m}$

difference in reading on the balance = $42.739 - 42.364 = 0.375\,\text{g}$

force between the spheres = $0.375 \times 10^{-3} \times 9.8 = 3.68 \times 10^{-3}\,\text{N}$

Using $F = \dfrac{Q_1 Q_2}{4\pi\varepsilon_0 r^2}$

$3.68 \times 10^{-3} = \dfrac{Q^2}{4\pi \times 8.85 \times 10^{-12} \times (6.2 \times 10^{-2})^2}$

$Q^2 = 3.68 \times 10^{-3} \times 4\pi \times 8.85 \times 10^{-12} \times (6.2 \times 10^{-2})^2 = 1.57 \times 10^{-15}\,\text{C}^2$

$Q = 4.0 \times 10^{-8}\,\text{C}$

You will remember that the electric field strength at a point is defined as the force per unit charge on a stationary positive point charge placed at that point. This means that the electric field strength around a point charge is given by the equation:

$$E = \dfrac{Q}{4\pi\varepsilon_0 r^2}$$

The equations for electric field strength and the force between two charged spheres should remind you of the equations for gravitational field and the gravitational force between two masses.

There is one important difference. Look at the two equations:

$F = \dfrac{Q_1 Q_2}{4\pi\varepsilon_0 r^2}$ and $F = -\dfrac{GMm}{r^2}$

You will observe that there is no minus sign in the equation for electric field strength. The minus sign indicates that forces in the gravitational field are always attractive. In an electric field the forces may be either attractive or repulsive. For repulsion, the sign is positive. We have already seen that:

- like charges repel
- the products of two positives (positive times positive) and of two negatives (negative times negative) are both positive

For attraction to occur, the charges must have opposite signs (positive and negative) so the negative sign comes in automatically where it is required.

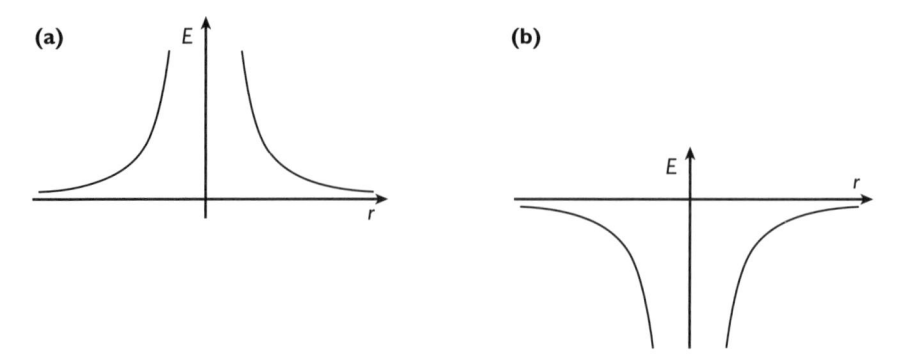

The electric field near (a) a positive point charge and (b) a negative point charge

Worked example
Calculate the electric field strength at a distance of 10 nm from an electron.

Answer

$$E = \frac{Q}{4\pi\varepsilon_0 r^2} = \frac{-1.6 \times 10^{-19}}{4\pi \times 8.85 \times 10^{-12} \times (10 \times 10^{-9})^2} = -1.4 \times 10^7 \, \text{N C}^{-1}$$

Electric potential

We saw in the work on gravity how gravitational potential at a point is defined as the work done in bringing unit mass from infinity to that point. When considering the electric field the rules are similar:
- Choose a point which is defined as the zero of electric potential — infinity.
- The electric potential at a point is then defined as the work done in bringing unit positive charge from infinity to that point.

Equation for the potential near a point charge:

$$V = \frac{Q}{4\pi\varepsilon_0 r}$$

Note the similarity with gravitational potential and, as with the electric field, there is no requirement for the minus sign. Although this equation refers to a point charge, it is, as with the gravitational example, a good approximation provided that:
- the charge is considered to be at the centre of the charged object and the distance is measured from this point
- the point considered is at a distance greater than the radius of the charged body

It follows from this that the electric potential energy of a charge Q_2 in an electric field is given by the equation:

$$E_P = \frac{Q_1 Q_2}{4\pi\varepsilon_0 r^2}$$

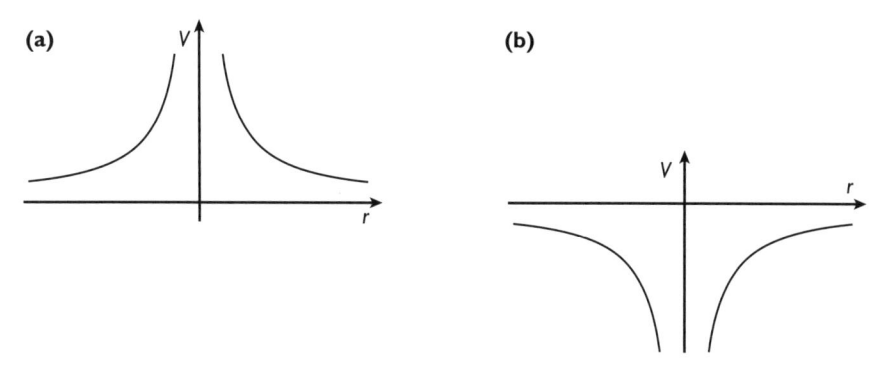

(a)

(b)

(a) When a positive charge is brought towards a positively charged body it gains potential energy. A 'potential hill' is formed.
(b) When a negative charge moves towards a positive charge, it loses potential energy. A 'potential well' is formed.

It is worth considering the potential energy of a positive charge and a negative charge when they are brought up to a positively charged body. Both charges have zero potential energy at infinity. As the positive charge is brought up to the body it gains potential energy. The negative charge loses energy as it approaches the body, so it has a negative amount of energy.

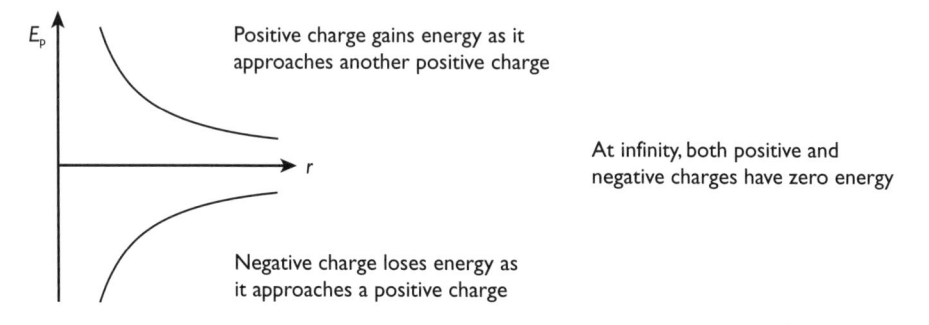

E_p

Positive charge gains energy as it approaches another positive charge

At infinity, both positive and negative charges have zero energy

Negative charge loses energy as it approaches a positive charge

This should help you understand why infinity is a good choice of position as the zero of potential energy.

Worked example

A proton travels directly towards the nucleus of an atom of silver at a speed of $5.00 \times 10^6 \, \text{m s}^{-1}$.

Calculate:
(a) the initial kinetic energy of the proton

(b) its closest approach to the silver nucleus

You may consider both the proton and the silver nucleus to be point charges.

(charge on the proton = +e, mass of a proton = 1.66×10^{-27} kg, charge on the silver nucleus = +47e, $e = 1.60 \times 10^{-19}$ C)

Answer
(a) $E_k = \tfrac{1}{2}mv^2 = 0.5 \times 1.66 \times 10^{-27} \times (5 \times 10^6)^2 = 2.08 \times 10^{-14}$ J

(b) Assume that all the potential energy of the proton is converted into electrical potential energy as it approaches the silver nucleus.

$$E_p = \frac{Q_1 Q_2}{4\pi\varepsilon_0 r^2}$$

$$2.08 \times 10^{-14} = \frac{47 \times 1.60 \times 10^{-19} \times 1.60 \times 10^{-19}}{4 \times \pi \times \varepsilon_0 \times r}$$

$$r = \frac{47 \times (1.60 \times 10^{-19})^2}{4 \times \pi \times 8.85 \times 10^{-12} \times 2.08 \times 10^{-14}}$$

$r = 5.20 \times 10^{-13}$ m

Relationship between electric field strength and potential

When studying the uniform field, we saw that the electric field strength between two plates can be calculated from either of the following two equations:

$E = F/q$ or $E = V/d$

The diagram below shows that in a uniform field the potential changes linearly with the distance d moved between the plates.

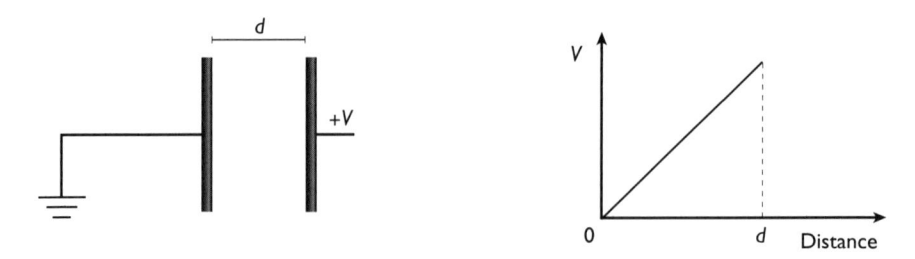

The second equation is valid only because the field between the plates is uniform and, hence, there is a steady change in potential from the earthed plate to the positive plate.

The field of a point charge gets weaker moving away from the charge. Consequently, the change in potential is not uniform. Nevertheless, the change in potential with respect to distance is equal to the gradient of the graph:

E = −gradient of the V–r graph

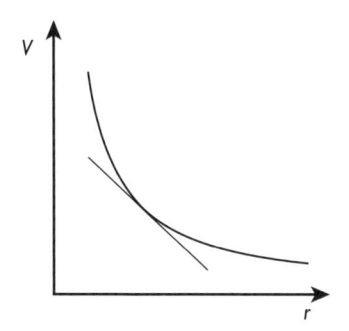

The electric field strength at any point is equal to minus the gradient at that point, which is equal to $-\Delta V/\Delta r$

The minus sign shows that the electric field (E) acts in the direction of decreasing potential. You can visualise this because when you bring a positive charge towards another positive charge you are going up a potential hill and the force is pushing you down the hill, in the opposite direction.

Capacitors

Capacitors are electronic devices that store charge. We can consider them to be made up of a pair of parallel conducting plates separated by an insulating material. When connected to a battery, charge flows onto one plate and an equal charge flows off the other plate.

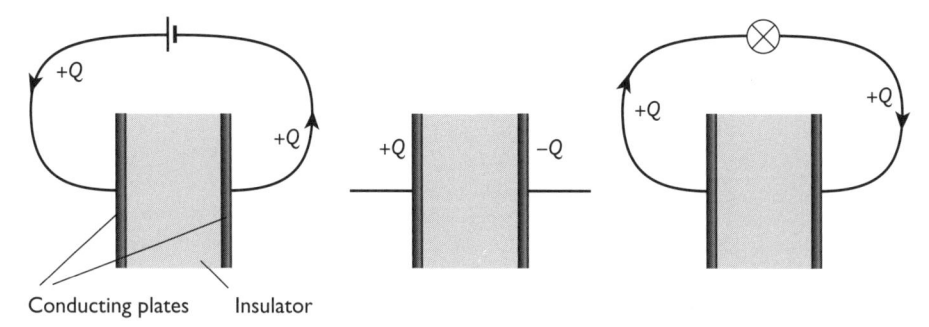

When the battery is removed there is a net positive charge on one plate and a net negative charge on the other. When the capacitor is connected to a light bulb it will light for a short time as the charge flows off the plates.

Capacitors have various uses in circuits; they are used in computers to run the computer for a long enough time to save data if there is a power cut. They are used to stop surges and to stop sparking when high voltages are switched. They can also be used as a time delay. For example, the timer on a burglar alarm system, which allows the operator time to leave the premises before the alarm switches on, will contain capacitors.

How much charge does a capacitor store?

This depends on the particular capacitor and the potential difference across it. The charge stored is proportional to the potential difference across the capacitor.

$Q \propto V$ can be rewritten as $Q = CV$ where C is a constant called the **capacitance** of the capacitor.

Capacitance is defined as the charge stored per unit potential difference across the capacitor.

$$C = Q/V$$

The unit of capacitance is the **farad (F)**.

One farad is the capacitance of a conductor that is at a potential of 1 V when it carries a charge of 1 C.

A capacitance of 1 F is huge. In general, the capacitance of capacitors in electronic circuits is measured in microfarads (μF) or picofarads (pF):

$1\,\mu F = 10^{-6}\,F$

$1\,pF = 10^{-12}\,F$

> **Worked example**
> Calculate the charge stored when a 2200 μF capacitor is connected across a 9 V battery.
>
> **Answer**
> $Q = CV = 2200 \times 10^{-6} \times 9 = 0.020\,C$ (20 mC)

It is not only capacitors that have capacitance — any isolated body has capacitance. Consider a conducting sphere of radius r carrying charge Q.

The potential at the surface of the sphere, $V = \dfrac{Q}{4\pi\varepsilon_0 r}$

$$C = \dfrac{Q}{V} = \dfrac{Q}{Q/4\pi\varepsilon_0 r}$$

Cancelling and rearranging gives:

$$C = 4\pi\varepsilon_0 r$$

> **Worked example**
> Estimate the capacitance of the Earth.
>
> (radius of the Earth = 6.4×10^6 m)
>
> **Answer**
> $C = 4\pi\varepsilon_0 r = 4 \times \pi \times 8.85 \times 10^{-12} \times 6.4 \times 10^6 = 7.1 \times 10^{-4}\,F \approx 700\,\mu F$

> **Note**
>
> This is surprisingly small and it shows just how big the unit 1 farad is. Remember that this is the capacitance of an isolated body. Practical capacitors have their plates very close together to increase their capacitance.

Capacitors in parallel

Consider the three capacitors in the diagram.

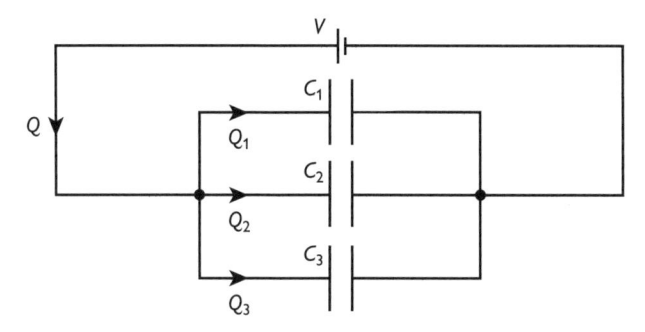

From Kirchhoff's second law, each capacitor has the same potential difference (V) across it.

From Kirchhoff's first law, the total charge flowing from the cell is the sum of the charges on each of the capacitors:

$Q = Q_1 + Q_2 + Q_3$

From $Q = CV$:

$C_{total}V = C_1V + C_2V + C_3V$

The Vs can be cancelled giving:

$C_{total} = C_1 + C_2 + C_3$

It is worth noting that whereas resistors in series add, it is capacitors in parallel that add. This is because by adding a capacitor in parallel we are providing an extra area for the charge to be stored on.

Capacitors in series

Consider three capacitors in series:

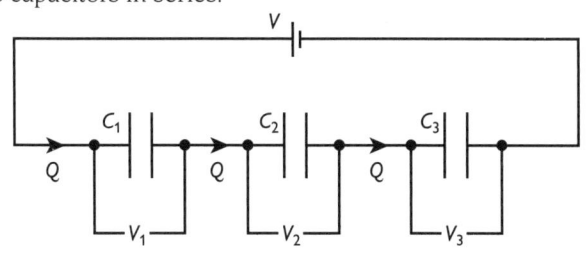

From Kirchhoff's second law, the potential difference across the cell will equal the sum of the potential differences across the capacitors:

$$V = V_1 + V_2 + V_3$$

But $Q = CV$, therefore $V = Q/C$, where C is the capacitance of the circuit, so:

$$Q/C = Q/C_1 + Q/C_2 + Q/C_3$$

The same charge Q flows on to each capacitor. Therefore the charges in the equation can be cancelled, so:

$$1/C = 1/C_1 + 1/C_2 + 1/C_3$$

This equation is of the same form as the formula for resistors in parallel. The combined resistance of several resistors in parallel is smaller than that of any of the individual resistors. Likewise, the combined capacitance of several capacitors in parallel is smaller than that of any of the individual capacitors.

Worked example

A student has three capacitors of values 47 µF, 100 µF and 220 µF.

(a) Calculate the total capacitance when:
  (i) all three capacitors are connected in parallel
  (ii) all three capacitors are connected in series

(b) How would the capacitors have to be connected to obtain a capacitance of 41 µF?

Answer

(a) (i) For capacitors in parallel:
  $C_{total} = C_1 + C_2 + C_3 = 47 + 100 + 220 = 367$ µF
  (ii) For capacitors in series:
  $1/C_{total} = 1/C_1 + 1/C_2 + 1/C_3 = 1/47 + 1/100 + 1/220 = 0.036$ µF^{-1}
  Therefore, $C = 28$ µF

(b) The capacitance is less than 47 µF, the value of the smallest of the three capacitances. Therefore, it is likely that the arrangement is of the form shown in the diagram:

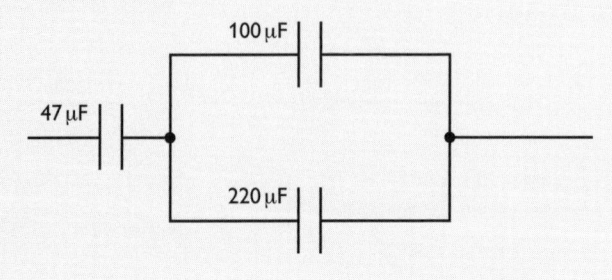

> Checking:
> capacitance of the two capacitors in parallel = 100 + 220 = 320 µF
> These are in series with the 47 µF capacitor so:
> $1/C_{total} = 1/C_1 + 1/C_2 = 1/47 + 1/320 = 1.244\,\mu F^{-1}$
> Therefore $C = 41\,\mu F$

Energy stored in a capacitor

Up to now we have described a capacitor as a charge store. It is more accurate to describe it as an energy store. The net charge on a capacitor is in fact zero: $+Q$ on one plate, $-Q$ on the other.

The energy stored in a capacitor is equal to the work done in charging the capacitor.

$W = ½QV$

The half comes in because:
- When the first charge flows onto the capacitor plates there is no potential difference opposing the flow.
- As more charge flows the potential difference increases, so more work is done.
- The average potential difference is equal to half the maximum potential difference.

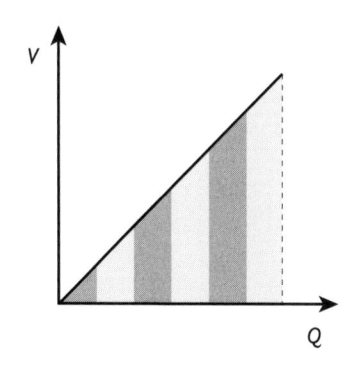

Each strip in the diagram shows the work done on adding a small amount of charge to a capacitor. It can be seen that the total work done in charging the capacitor, and hence the energy it stores, is equal to the area under the graph.

It is sometimes useful to express the energy equation in terms of capacitance and potential difference. Substitute for Q using the basic capacitor equation, $Q = CV$:

$W = ½QV = ½(CV)V$

Therefore:

$W = ½CV^2$

Worked example 1
Calculate the energy stored in a 470 µF capacitor when it is charged with a 12 V battery.

Answer
$W = \frac{1}{2}CV^2 = 0.5 \times (470 \times 10^{-6}) \times 12^2 = 3.4 \times 10^{-2}$ J

Worked example 2
A 50 µF capacitor is charged using a 12 V battery. It is then disconnected from the battery and connected across a second 50 µF capacitor.

Calculate:
(a) the charge on the first capacitor before it is connected to the second capacitor

(b) the energy stored on the first capacitor

(c) the charge on each capacitor when they are connected together

(d) the potential difference across the capacitors

(e) the total energy stored when the two capacitors are connected together

Answer
(a) $Q = CV = 50 \times 12 = 600$ µC $= 6.0 \times 10^{-4}$ C

(b) $W = \frac{1}{2}QV = 0.5 \times 6 \times 10^{-4} \times 12 = 3.6 \times 10^{-3}$ J $= 3.6$ mJ

(c) The charge will be shared equally between the two capacitors; therefore each capacitor has a charge of 3.0×10^{-4} C.

(d) The charge on each capacitor is half the original value, so the p.d. will be one-half of the original = 6.0 V

(e) Energy on one of the capacitors = $\frac{1}{2}QV = 0.5 \times 3.0 \times 10^{-4} \times 6 = 9.0 \times 10^{-4}$ J

Therefore the total energy = $2 \times 9.0 \times 10^{-4} = 1.8 \times 10^{-3}$ J

Note

You will observe that in Worked example 2 above the answer is half the original energy. The remainder is dissipated in the wires as work is done against their resistance. Another way of tackling this problem would have been to consider the two capacitors as capacitors in parallel.

Magnetic fields

You should be familiar with magnetic fields from your earlier work. This section reviews that work in preparation for studying currents in magnetic fields.

Magnetic effects

Iron, cobalt and nickel and many of their alloys show **ferromagnetic effects**. Magnets are generally made of steel, an alloy of iron and carbon that retains magnetism much better than iron alone. The ends of a magnet are called poles. One end is the north-seeking pole and the other is the south-seeking pole (usually shortened to north and south poles).

The laws of magnetism are that:
- like poles repel
- unlike poles attract

Magnetic field shapes

Just as with gravitational fields and electric fields, a magnetic field is a region in which a force is felt. In this case, the force is the force on a single north pole placed in the field. The field shape can be shown using lines of magnetic force, which are known as **magnetic field lines.** These are imaginary lines that show the direction of the force on a free north pole when it is placed in the field. Just as with electric and gravitational field diagrams, the stronger the field, the closer the lines are drawn.

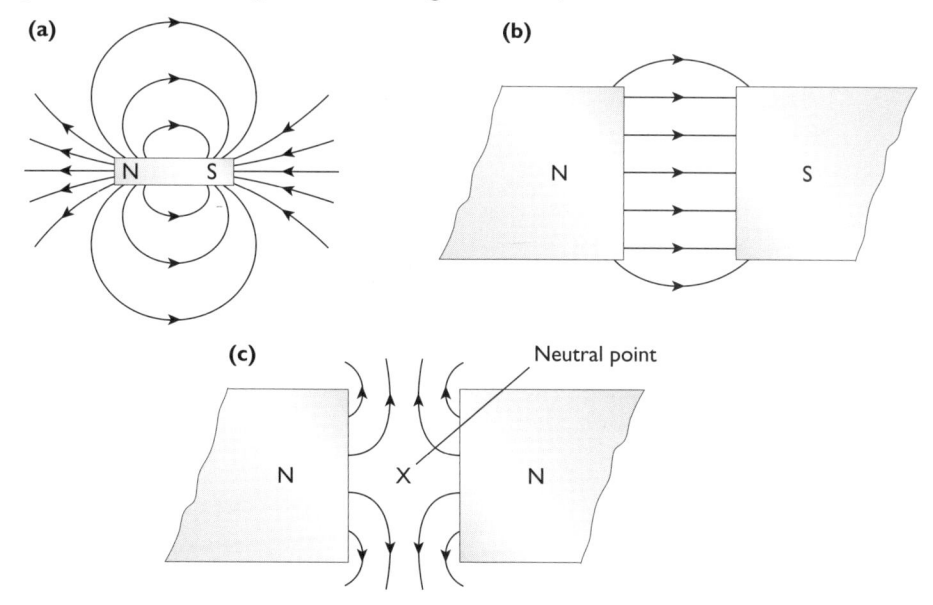

(a) Magnetic field of a bar magnet
(b) Magnetic field between opposite poles
(c) Magnetic field between like poles. At the neutral point the two fields cancel out and there will be no force on a single north pole situated at this point

From the diagram above you can see that the lines of magnetic field start at a north pole and finish on a south pole, and that they never cross or touch.

V Electricity and magnetism

Magnetic field of an electric current

It is not only magnets that have an associated magnetic field — currents also do. Indeed, the origin of the magnetic field in a magnet is due to circulating charges (currents) in the atoms that make up the magnet.

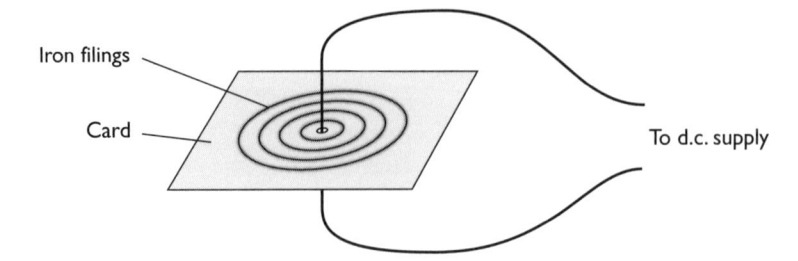

The magnetic field of a straight current-carrying conductor can be investigated using the apparatus shown in the diagram above.

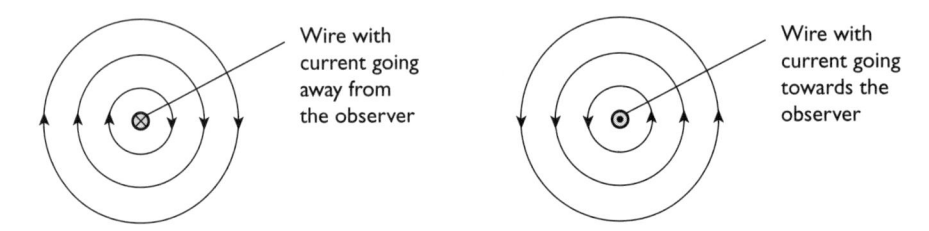

The field of a straight current-carrying conductor is a set of concentric circles. The direction of the field depends on the current direction. You can work out the direction by using the screw rule, 'Imagine you are screwing a screw into the paper. The screw driver must turn clockwise, the same direction as the field lines for a current going into the plane of the paper'.

The magnetic fields of pairs of conductors with the currents in the opposite and same directions are shown in the diagram below.

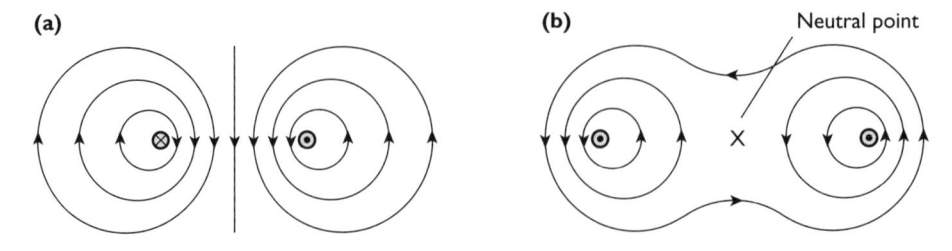

(a) The magnetic field of a pair of conductors with the currents in opposite directions. Note that this also shows the field of a narrow coil viewed at 90° to the axis of the coil

(b) The magnetic field of a pair of conductors with the currents in the same direction. Note the neutral point where the fields cancel each other out

The magnetic field of a current-carrying solenoid is shown in the diagram below.

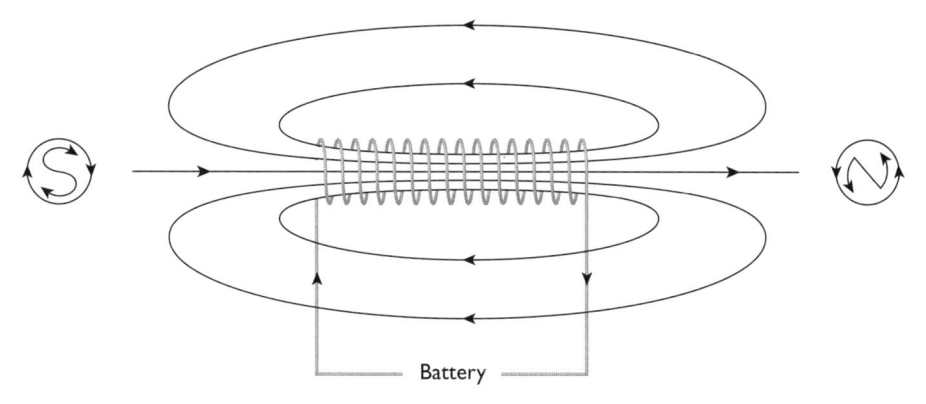

You should note that:
- the field is very strong inside the solenoid
- the field is similar to that of a bar magnet

The circles containing the N and S show a way of remembering which end of the solenoid acts as which pole. If you look at the right-hand end of the solenoid the current direction is anticlockwise, the same as the arrows on the N. Look at the left-hand end. The current is in a clockwise direction, the same as the arrows on the S.

Solenoids are often used to make electromagnets, with the solenoid being wound round an iron former. The presence of the iron greatly increases the strength of the magnetic field inside (and near) the solenoid.

Electromagnetism

Force on a current-carrying conductor: the motor effect

When a current-carrying conductor lies in a magnetic field there is a force on the conductor. This is called the **motor effect**.

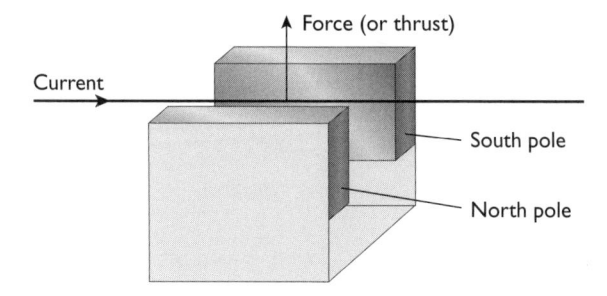

You will observe that the current and the field are at right angles to each other, and that the force is at right angles to both of these, thus making a set of three-dimensional axes. There will be no force on the conductor if it is parallel to the field, it requires at least a component of the current at right angles to it.

To help remember the specific directions of the different vectors we use **Fleming's left-hand rule**, in which the **fi**rst finger represents the **fi**eld, the se**c**ond finger the **c**urrent and the **th**umb the **th**rust (or force):

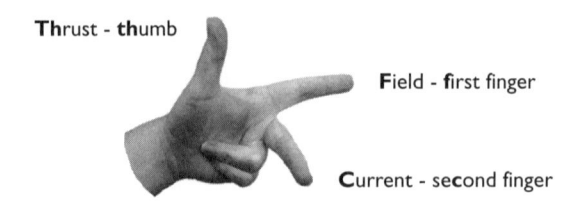

Fleming's left-hand rule

The magnitude of the force depends on:
- the size of the current
- the length of conductor in the field
- the angle the conductor makes with the field

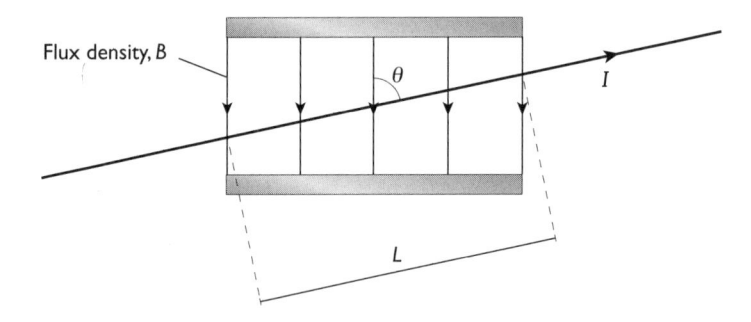

The force on a current carrying conductor at an angle θ to a magnetic field

A conductor of length L carrying a current I, lying at an angle θ to the field will experience a force F vertically into the plane of the paper.

$$F \propto IL\sin\theta$$

This can be written as:

$F = BIL\sin\theta$

where B is a constant, which can be considered as the **magnetic field strength**, although for historical reasons it is more usual to call it the **flux density**.

Flux density is defined from the rearranged equation:

$$B = \frac{F}{IL\sin\theta}$$

Flux density is numerically equal to the force per unit length on a straight conductor carrying unit current at right angles to the field.

The units of flux density are $N\,A^{-1}\,m^{-1}$; $1\,N\,A^{-1}\,m^{-1}$ is called **1 tesla** (T).

1 tesla is defined as the magnetic flux density that, acting normally to a straight conductor carrying a current of 1 A, causes a force per unit length of 1 N m^{-1}.

> **Worked example**
> A copper power cable of diameter 2.5 cm carries a current of 2000 A to a farm. There is a distance of 50 m between successive telegraph poles.
>
> (a) Calculate the magnetic force on the section of the cable between two telegraph poles due to the Earth's magnetic field. (You may consider the wire to be at right angles to the Earth's magnetic field.)
>
> (b) Compare this with the gravitational force on the cable.
>
> (density of copper = 8900 kg m^{-3}, the flux density of the Earth's field = 30 µT)
>
> **Answer**
> (a) $F = BIL \sin\theta = 30 \times 10^{-6} \times 2000 \times 50 \times \sin 90 = 3.0$ N
>
> (b) volume of the copper wire = $\pi r^2 L = \pi \times (2.5 \times 10^{-2}/2)^2 \times 50 = 0.025$ m^3
>
> mass = density × volume = 8900 × 0.025 = 220 kg
>
> weight = mg = 220 × 9.8 ≈ 2200 N
>
> This is almost three orders of magnitude (i.e. ×1000) greater than the magnetic force on the cable.

Measurement of flux density

The flux density can be investigated using a simple current balance.

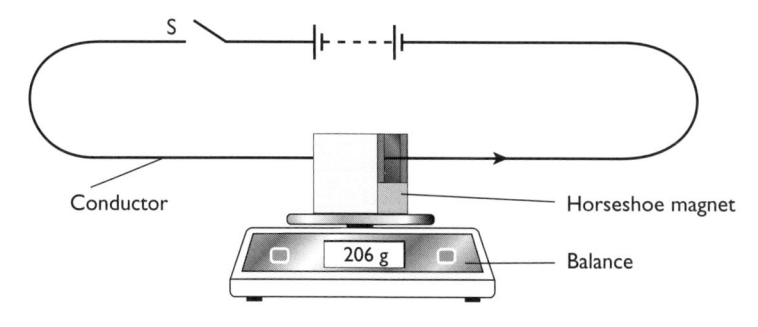

A current balance

The current balance relies on Newton's third law. Not only does the magnetic field cause a force on the current, the current causes an equal-sized force on the magnets but in the opposite direction. So the difference in the readings on the top-pan balance when switch S is open and closed gives a measure of the electromagnetic force.

Worked example
A student uses a current balance to measure the magnetic flux density of a horseshoe magnet. He measures the length of the pole pieces (8.1 cm) and places the magnet on the top-pan balance, which reads 95.452 g. He sets up the current balance as shown in the diagram on page 185 and closes the switch. The ammeter reading is 2.3 A, and the reading on the top-pan balance falls to 95.347 g.

(a) Calculate the flux density between the poles of the magnets.

(b) Discuss the effect on the result if the magnetic pole pieces are not exactly parallel to the wire.

(c) Discuss one other factor that may limit the accuracy of the result.

Answer
(a) change in the balance reading = 95.452 − 95.347 = 0.105 g

force on the magnets = 1.03×10^{-3} N

$$B = \frac{F}{IL \sin \theta} = \frac{1.03 \times 10^{-3}}{2.3 \times 8.1 \times 10^{-2} \times \sin 90} = 5.5 \times 10^{-3} \text{ T}$$

(b) It will have no effect, because although the force per unit length of wire would be reduced by a factor of $\sin \theta$, where θ is the angle the wire makes with the magnetic field lines, the length of wire between the poles will increase by the same factor. Thus the two changes cancel each other out.

(c) There is likely to be some fringing of the field at the edges of the poles thus increasing the length of wire in the field. However, this might be reduced because the field at the edges of the pole pieces is weaker than near the centre. In practice, this system only measures a mean (or average) field.

Forces on parallel current-carrying conductors

You are already familiar with the fact that a current has an associated magnetic field.

The diagram below shows parallel conductors with currents in opposite directions. It is a model in which the fields are considered separately. In practice, the two fields will combine.

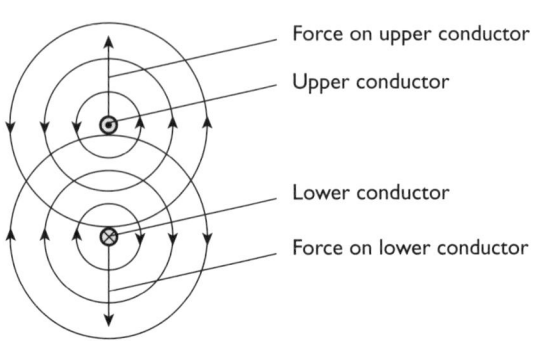

Consider the effect of the field of the lower conductor on the upper conductor. The magnetic field of the lower conductor is to the right of the page and the current in the upper conductor is vertically upwards, out of the plane of the paper. If you apply Fleming's left-hand rule you will see that there will be a force on the upper conductor away from the lower conductor.

If you consider the effect of the upper conductor on the lower conductor you will see that the force is vertically down the page. The two conductors repel.

A similar analysis shows that if the currents are both in the same direction then the two conductors will attract.

> **Note**
>
> It is worth noting that the ampere is defined in terms of the force per unit length between two parallel current-carrying conductors.

Forces on charged particles in a magnetic field

Electric current is a flow of electric charge, so the magnetic force on a current is the sum of the forces on all the moving charge carriers that make up the current. Alternatively, we could think of a beam of charged particles as a current. Either way we can deduce that the force on a charge q moving through a field of flux density B at speed v is given by the formula:

$$F = Bqv \sin \theta$$

where θ is the angle the velocity makes with the field.

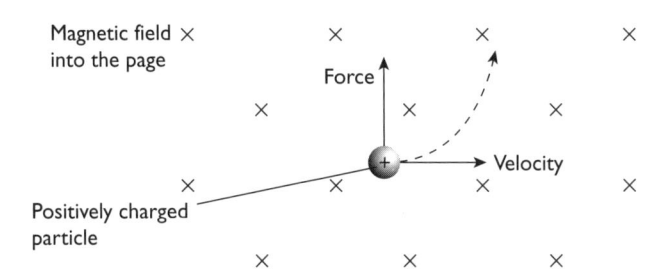

The path of a charged particle travelling with a velocity at right angles to the magnetic field

The direction of the force on the particle can be found by using Fleming's left-hand rule. Remember that the second finger shows the direction of the conventional current. Thus the thumb shows the force direction on a positive charge; a negative charge will experience a force in the opposite direction.

Study the diagram above. The force is at right angles to both the field and the velocity. As the velocity changes, so does the direction of the force. Consequently, the particle travels in a circular path with the magnetic force providing the centripetal force.

V Electricity and magnetism

You will study the tracks of charged particles in more detail in the section on charged particles (pages 201–206).

Comparison of forces in gravitational, electrical and magnetic fields

Table 19 shows the differences and similarities between the three types of force field we have met.

Table 19

	Gravitational	Electric	Magnetic
Stationary mass	Attractive force parallel to the field	None	None
Moving mass	Attractive force parallel to the field	None	None
Stationary charge	None	Attractive or repulsive force depending on type of charge, parallel to the field	None
Moving charge and electric current	None	Attractive or repulsive force depending on type of charge, parallel to the field	Force at right angles to both the field and the velocity of the charge/current. Force is a maximum when the velocity is at right angles to the field/current, and zero when the velocity/current is parallel to the field

Electromagnetic induction

Moving a wire perpendicularly to a magnetic field, as in the diagram below, induces an e.m.f. across the ends of the wire.

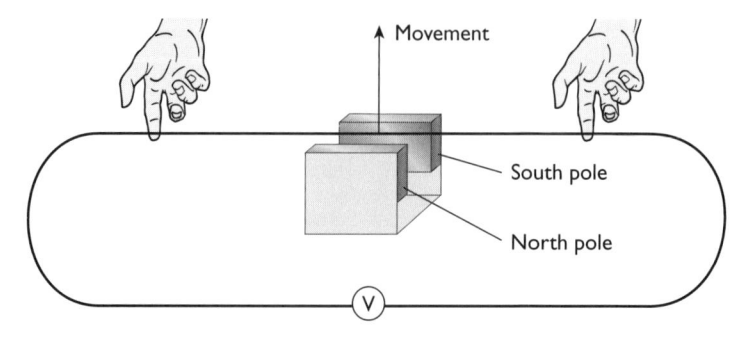

The magnitude of the e.m.f. depends on the strength of the magnet, the length of wire in the field and the speed at which it is moved. The direction of the e.m.f. depends on the direction in which the wire is moved.

If a second loop is made in the wire the induced e.m.f. is doubled.

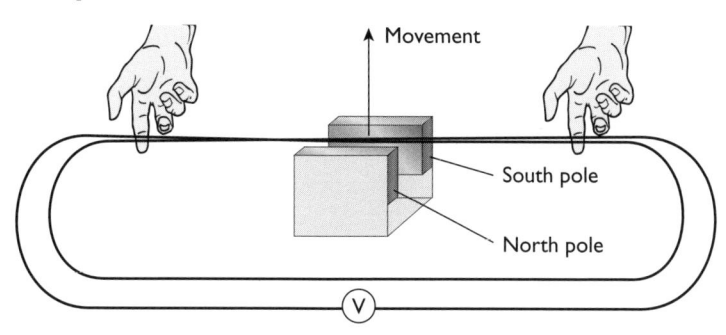

Further loops show that the induced e.m.f. is proportional to the number of loops, N. In practice, most electromagnetic devices consist of a coil of many turns of wire rather than just a single wire.

Magnetic flux

A wire moving through a magnetic field sweeps out an area, A, as shown in the diagram below.

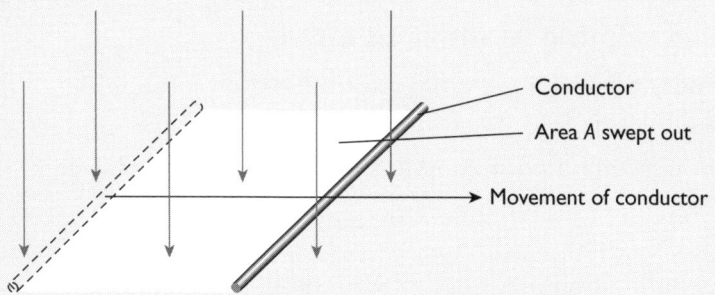

The flux density multiplied by this area (BA) is called the **magnetic flux** and has the symbol ϕ.

The unit of ϕ is the weber (Wb).

$1 \text{ Wb} = 1 \text{ T m}^2$.

Magnetic flux is the product of magnetic flux density and the area, normal to the field, through which the field is passing.

This leads to the following equation for the induced e.m.f. E:

$$E = -N \frac{\Delta \phi}{\Delta t}$$

This is often written:

$$E = -\frac{\Delta(N\phi)}{\Delta t}$$

$N\phi$ is called the magnetic **flux linkage**.

The magnetic flux linkage of a coil is the product of the magnetic flux passing through a coil and the number of turns on the coil.

Worked example

A small coil of cross-sectional area 2.4 cm² has 50 turns. It is placed in a magnetic field of flux density 4.0 mT, so that the flux is perpendicular to the plane of the coil.

The coil is pulled out of the field in a time of 0.25 s. Calculate the average e.m.f. that is induced in the coil.

Answer

The initial flux linkage $(N\phi) = NBA = 50 \times 4 \times 10^{-3} \times 2.4 \times 10^{-4} = 4.8 \times 10^{-5}$ Wb

$$E = -\frac{\Delta\phi}{\Delta t} = -\frac{4.8 \times 10^{-5}}{0.25} = -1.9 \times 10^{-4} \text{V}$$

Note

$1 \text{Wb s}^{-1} = 1 \text{V}$

Laws relating to induced e.m.f.

Faraday's law of electromagnetic induction

This law is:

The induced e.m.f. is proportional to the rate of change of magnetic flux linkage.

Notice that it is the *rate of change* of flux linkage, not just cutting through flux linkage: if a wire is in a magnetic field which changes, an e.m.f. is induced just as though the wire had been moved. The wire or coil 'sees' a disappearing magnetic flux.

Lenz's law

This law is really a statement of the conservation of energy. When a current is induced in a conductor, that current is in a magnetic field. Therefore, there is a force on it due to the motor effect. Work must be done against this force in order to drive the current through the circuit.

A formal statement of Lenz's law is as follows:

The direction of the induced e.m.f. is always in such a direction so as to produce effects to oppose the change that is causing it.

Worked example

Electromagnetic braking is used in trains and large vehicles. The diagram shows the principles of the system. To brake, the electromagnets are switched on, inducing eddy currents in the thick aluminium disc.

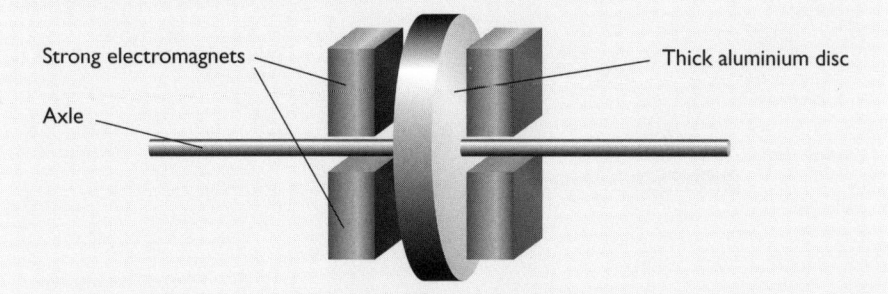

A large current is passed through the coils of the electromagnets to produce a uniform magnetic field across the disc of 0.80 T. The disc has a radius of 40 cm and it rotates at 24 revolutions per second.

Calculate the e.m.f. induced in the disc.

Answer

The area swept out per unit time by the disc = $\pi r^2 \times n$ where n = number of revolutions per second.

area = $\pi \times 0.40^2 \times 24 = 12.1 \, m^2$

induced e.m.f. $E = -\frac{\Delta(N\phi)}{\Delta t}$, where $N = 1$ and $\Delta\phi/\Delta t = B \times$ area swept out per second.

Therefore:

$E = 0.80 \times \pi \times 0.40^2 \times 24 = 9.7 \, V$

Note

This example could form the basis of an experiment to demonstrate Lenz's law. A conducting disc is spun round so that it rotates freely. It will slow down gradually due to the frictional forces at the bearings. If the experiment is repeated so that the disc is between the jaws of a strong magnet (as described in the example above) the disc will slow down much more quickly. The extra decelerating forces are produced by the eddy currents induced in the disc.

The Hall effect and Hall probe

The diagram shows a thin slice of a semiconductor.

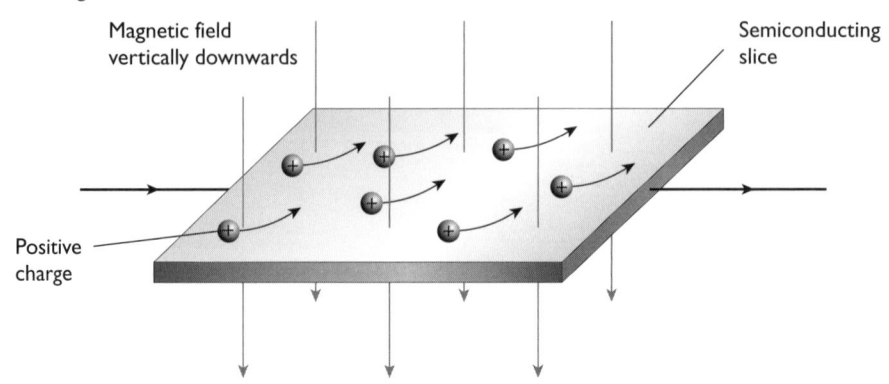

In this example, the current is carried by positive charge carriers. When a magnetic field is applied the charge carriers are pushed to the rear side of the slice. This produces an e.m.f. between the front and rear edges of the slice, known as the Hall voltage (V_H). The greater the input current, or the greater the magnetic flux density, the greater is the Hall voltage. This variation in voltage with field strength allows the effect to be used to measure magnetic flux density.

The Hall voltage produced in a conductor is much less than in a semiconductor. Therefore this device is always made from a semiconducting material.

When using a Hall probe care must be taken that the semiconductor slice is at right angles to the field being investigated.

Alternating currents

Terminology

Up to this point we have only looked at direct currents, which have been considered as steady unchanging currents. An alternating current changes direction continuously — the charge carriers vibrate backwards and forwards in the circuit.

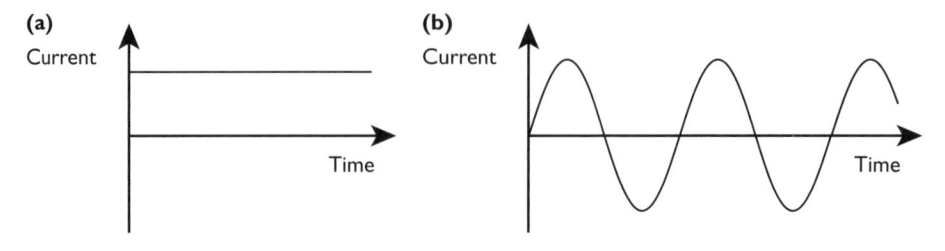

(a) A direct current from a battery
(b) An alternating current from a simple generator

The terminology used in this section is similar to that used in the work on oscillations.

The **frequency** (*f*) of the signal is the number of complete oscillations of the signal per unit time.

The **period** (*T*) is the time taken for one complete oscillation of the signal.

The **peak current** (I_0) is equal to the maximum current during the cycle.

This type of current is driven by an alternating voltage with a similar shaped curve. You should recognise the shape of the curve from the work you have done on oscillations. The equations for these curves are as follows:
- for the current $\quad I = I_0 \sin \omega t$
- for the voltage $\quad V = V_0 \sin \omega t$

where I_0 is the peak current, V_0 is the peak voltage, and ω is the angular frequency of the signal (= $2\pi f$).

> **Hint**
>
> Any current that changes direction continuously is described as an alternating current. In this course you need only concern yourself with sinusoidal alternating currents.

Power dissipated by an alternating current

Just as the current is changing continuously, the power dissipated is changing continuously.

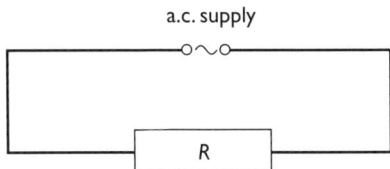

In the diagram, an a.c. supply is driving a current, *I*, which is equal to $I_0 \sin \omega t$, through a resistor of resistance *R*.

The power *P* dissipated by the alternating current in the resistor is:

$P = I^2 R$

$P = (I_0 \sin \omega t)^2 R$

$P = I_0^2 R \sin^2 \omega t$

V Electricity and magnetism

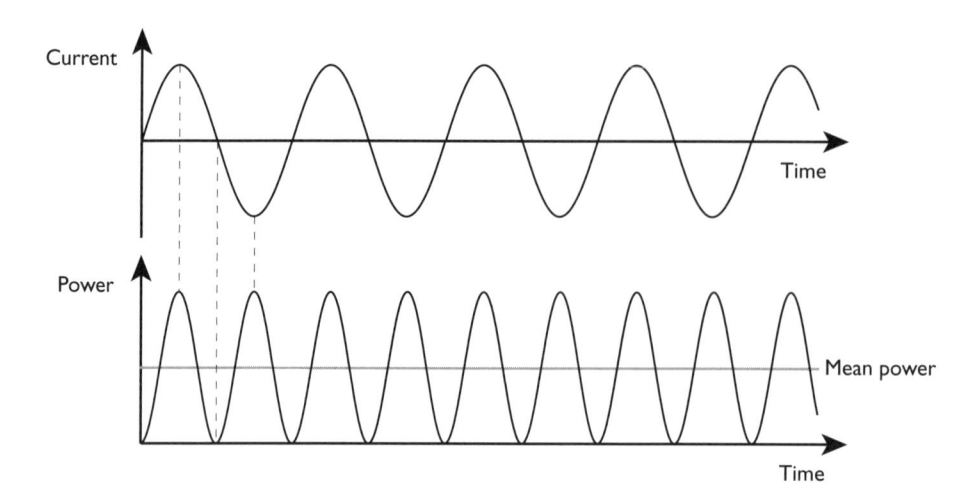

Graphs of the alternating current through a resistor and the power dissipated in the resistor

Study the power curve — it is always positive. Even though the current goes negative, power is equal to current squared, and the square of a negative number is positive.

The average power dissipated over a complete cycle is equal to half the peak power during that cycle (the grey line on the graph).

$$P_{average} = \tfrac{1}{2}P_0 = \tfrac{1}{2}I_0^2 R$$

The direct current that would dissipate this power = $\sqrt{\tfrac{1}{2}I_0^2}$.

$$I_{d.c.} = \frac{I_0}{\sqrt{2}}$$

This current is known as the root-mean square current (r.m.s.) current.

$$I_{r.m.s.} = \frac{I_0}{\sqrt{2}}$$

Similarly, the r.m.s. voltage is given by:

$$V_{r.m.s.} = \frac{V_0}{\sqrt{2}}$$

The r.m.s. value of the current (or voltage) is the value of direct current (or voltage) that would convert energy at the same rate in a resistor.

Worked example

(a) Explain what is meant by the statement that the mains voltage is rated at 230 V, 50 Hz. Calculate the peak voltage.

(b) Calculate the energy dissipated when an electric fire of resistance 25 Ω is run from the supply for 1 hour.

Answer
(a) 230 V tells us that the r.m.s. voltage is 230 volts. The frequency of the mains supply is 50 Hz.

$V_0 = V_{r.m.s.} \times \sqrt{2} = 230 \times \sqrt{2} = 325\,V$

(b) energy $= \dfrac{V_{r.m.s.}^2 t}{R} = \dfrac{230^2 \times 60 \times 60}{25} = 7.6 \times 10^6\,J$

Note

The r.m.s voltage is used, not the peak voltage.

The transformer

Transformers are used to step voltages up (e.g. for the accelerating voltages in a cathode-ray tube) or down (e.g. for an electric train set).

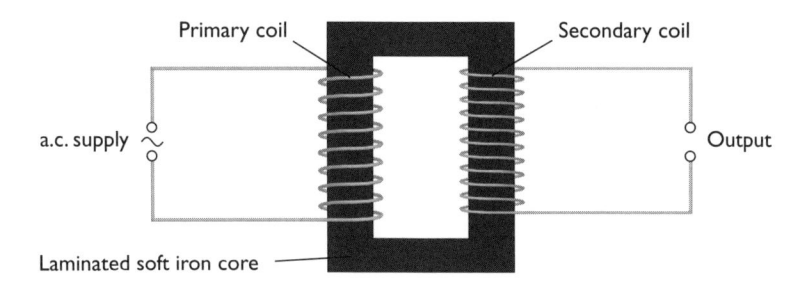

A transformer

A transformer works on the following principles:
- The alternating current in the primary coil produces an alternating magnetic flux in the soft iron former.
- The soft iron core strengthens the magnetic field produced by the current in the primary coil.
- The alternating flux in the transformer is transmitted round the core and cuts the secondary coil.
- The changing magnetic flux in the secondary coil induces an alternating e.m.f. across the ends of the secondary coil.

In an ideal transformer, when no current is taken from it:

$$\dfrac{V_s}{V_p} = \dfrac{N_s}{N_p}$$

where V_s is the e.m.f induced across the secondary coil, V_p is the voltage across the primary coil, N_s is the number of turns in the secondary coil and N_p is the number of turns in the primary coil.

V Electricity and magnetism

You can see from this that in a step-up transformer the primary coil will have only a few turns, while the secondary coil will have many turns. The reverse is true for a step-down transformer.

Worked example
An electric train set is designed to operate at 12 V a.c.

Calculate the turns ratio for a transformer that would be suitable to step down a mains voltage of 230 V.

Answer

$$\frac{V_s}{V_p} = \frac{N_s}{N_p} = \frac{12}{230} = 1:19$$

Power output from a transformer
In an ideal transformer the power output would equal the power input. Thus substituting for power:

$$V_s I_s = V_p I_p$$

Worked example
A 12 V, 96 W heater is run from a transformer connected to the 230 V mains supply.

Assuming that there are no energy losses in the transformer, calculate:
(a) the current in the heater
(b) the current input to the transformer

Answer
(a) $P = V_s I_s$

$I_s = P/V_s = 96/12 = 8.0$ A

(b) $V_s I_s = V_p I_p$

$12 \times 8 = 230 \times I_p$

Therefore $I_p = 12 \times 8/230 = 0.42$ A

Real transformers
In practice, transformers, although they can be designed to have efficiencies in excess of 99%, are never 100% efficient. Energy losses come from:
- work done in overcoming the resistance of the coils; the coils are usually made from copper and hence these energy losses are known as copper losses
- the induction of currents in the iron former (known as eddy currents); these energy losses are known as iron losses

- work done in the formation of magnetic fields in the iron core; these energy losses are known as hysteresis losses

The energy losses due to the resistance of the coils are kept to a minimum by making the coils out of a good conductor, such as copper.

In a step-up transformer the input current is large, so the few turns required are made from thick wire. The secondary coil, which carries only a small current, is made from much thinner wire.

The iron losses are reduced by laminating the core. This means making it out of thin iron plates, each plate being insulated from its neighbours by a thin layer of varnish. This reduces the size of the eddy currents induced.

Hysteresis losses are reduced by making the core from pure iron, which is easily magnetised and demagnetised, rather than from steel which requires much more work to magnetise and demagnetise.

Transmission of electrical energy

Electrical energy is transferred from power stations across many kilometres of power lines to local communities. Although the cables are thick and made from metals of low electrical resistance, energy is still dissipated in them.

energy dissipated in the cable = I^2R, where R is the resistance of the cable

The smaller the current, the less energy is wasted in the cable. Indeed, if the current is halved then the power loss is reduced by a factor of four. Consequently, energy is transmitted at high voltage–low current and is then stepped down before being distributed to the consumer. Alternating currents are used for transmitting electrical energy because the alternating voltages can be stepped up and down with much less energy loss than direct voltages.

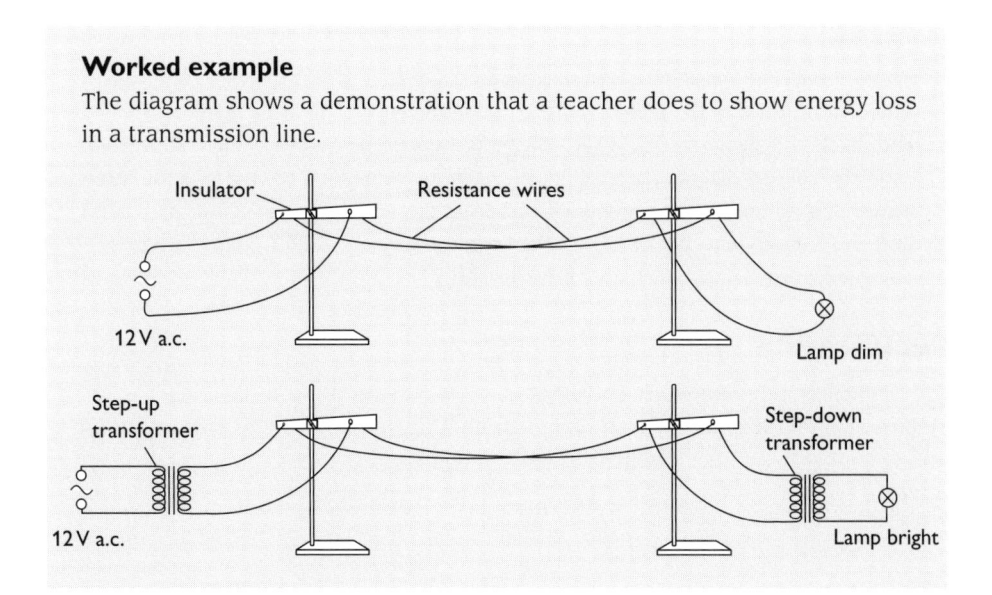

Worked example
The diagram shows a demonstration that a teacher does to show energy loss in a transmission line.

V Electricity and magnetism

The two resistance wires have a total resistance of 2.0 Ω. The lamp is designed to run at 12.0 V and to transfer energy at the rate of 36 W.

(a) Calculate the resistance of the lamp and the total resistance in the circuit. (Assume that the resistance of the lamp does not change with temperature.)

(b) Calculate the current in the circuit.

(c) Calculate the drop in potential across the resistance wires.

(d) Calculate the power loss in the resistance wires and the power dissipated in the lamp.

Answer

(a) $P = V^2/R$

therefore $R = V^2/P = 12^2/36 = 4.0\,\Omega$

total resistance in the circuit = 2.0 + 4.0 = 6.0 Ω

(b) $I = V/R = 12/6.0 = 2.0\,A$

(c) potential drop across the wires = resistance of the wires × current through them = 2.0 × 2.0 = 4.0 V

(d) power dissipated in the wires = potential drop across wires × current = 4.0 × 2.0 = 8.0 W

(e) power dissipated in the lamp = potential drop across lamp × current = (12.0 − 4.0) × 2.0 = 16.0 W

Note

You will see that one-third of the total potential drop is across the wires and consequently one-third of the power is dissipated in the wires. The lamp lights only dimly and much energy is wasted. If the voltage is stepped up before being transmitted, as in the second diagram, a much smaller current is needed to transmit the same power. Consequently the potential drop across the wires (and the power dissipated in them) is much less. The voltage is then stepped down to 12.0 V once more before being fed to the lamp, which now lights brightly.

Rectification

Although it is advantageous to transmit power using alternating currents, many electrical appliances require a direct current. The simplest way to convert an alternating supply to a direct supply is to use a single diode.

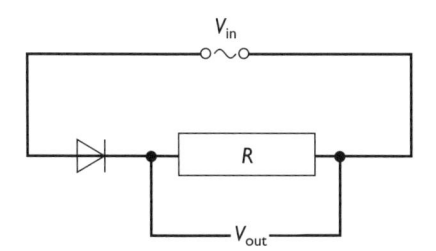

The diode allows a current to pass only one way through it. The current through the resistor causes a potential drop across it when the diode conducts. When the input voltage is in the opposite direction, there is no current through the resistor, so the potential difference across it is zero. This is known as **half-wave rectification**.

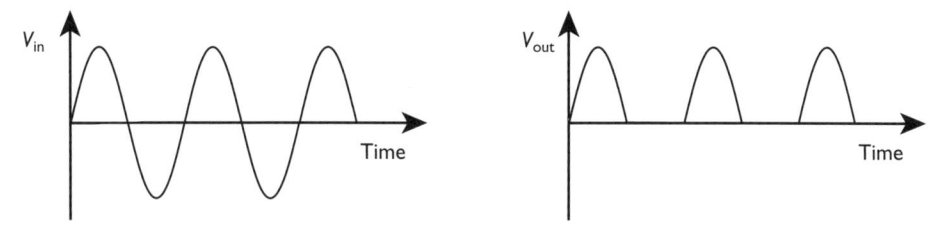

Graph showing half-wave rectification

Full-wave rectification

With half-wave rectification there is a current for only half a cycle. To achieve full-wave rectification, a diode bridge is used.

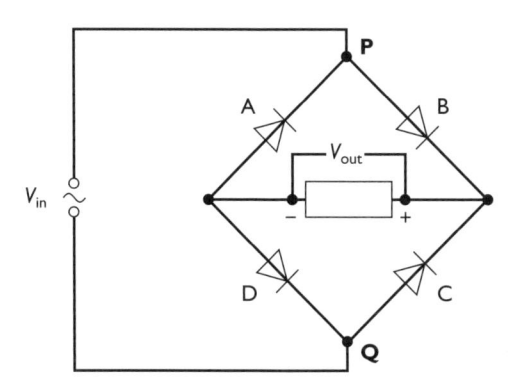

Diode bridge for full-wave rectification

If you study the diagram above, you can see that when point P is positive with respect to Q, then a current will pass from P through diode B, through the resistor, then through diode D to point Q. When point P is negative with respect to Q, then the current will pass from Q through diode C, through the resistor, and through diode A to point P.

In both cases the current is in the same direction through the resistor, so the potential difference across it is always in the same direction. There is a full-wave rectified output voltage, as shown in the graph below.

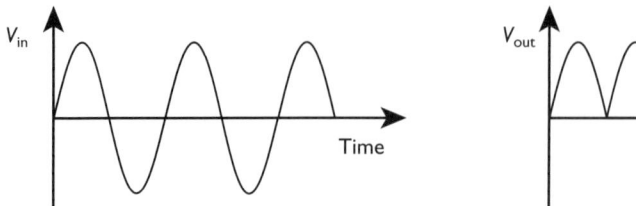

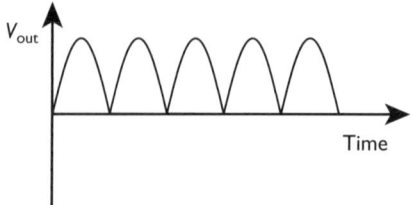

Smoothing

The output from a full-wave rectifier is still rough, rising from zero to a maximum and back to zero every half cycle of the original alternating input. Many devices, such as battery chargers, require a smoother direct current for effective operation. To achieve this, a capacitor is connected across the output resistor.

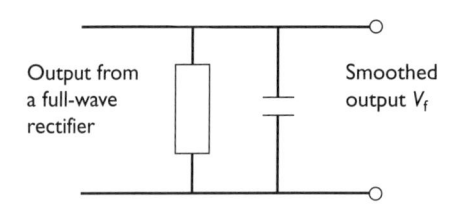

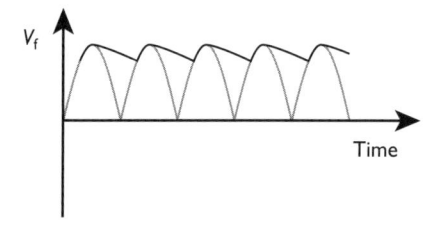

Smoothing circuit and the smoothed output produced by using a single capacitor

Because the capacitor takes some time to discharge, it will only partially discharge in the time it takes for the potential difference to rise once more. The value of the product *CR* (where *C* is the capacitance of the capacitor and *R* is the load resistance) should be much greater than the time period of the original alternating input. This means that the capacitor does not have sufficient time to discharge significantly.

Note

The unit of capacitance is the farad; the unit of resistance is the ohm.

$F = CV = As V^{-1}$ and $\Omega = V A^{-1}$

So the units of $CR = F\Omega = As V^{-1} \times V A^{-1} = s$

The product *CR* is called the time constant of a capacitor–resistor circuit. It is the time taken for the charge on the capacitor to fall to 1/e times the original value.

You will meet something similar to this when you study the decay of a radioactive isotope. It is worth noting that increasing either the resistance or the capacitance will decrease the 'ripple' on the output voltage.

Modern physics
Charged particles

Measurement of the charge on an electron

The measurement of the charge on an electron gave evidence for the quantisation of charge. It showed that there is a fundamental charge that cannot be split further.

(Quantisation means that the charge is in discrete amounts.)

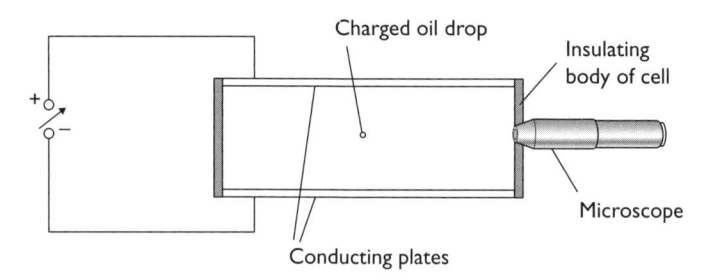

The Millikan experiment to measure the charge on the electron

Droplets of oil were sprayed between two charged plates a distance d apart and at a potential difference of V. The weight of a tiny droplet of oil of mass m carrying a charge q was balanced by the electrical force due to the electric field E.

$$Eq = (V/d)q = mg$$

where g is the acceleration due to gravity.

The oil drops were so small that the charge on them was caused by an excess or shortage of only a few electrons. A drop was observed and balanced by adjusting the voltage across the plates. This was repeated for many drops. From all the results, Millikan observed that the charge on the oil drop was always an integral multiple of 1.6×10^{-19} C. He concluded that this was the charge carried on a single electron. Hence he inferred that charge was quantised.

Deflection of charged particles in fields

Electric fields
Accelerating field

An electric field can be used to accelerate a charged particle in the direction of the field. Early particle accelerators used electrostatic fields in this way. The kinetic energy given to the particle carrying charge q is:

$$E_k = Vq$$

Therefore:

$$\tfrac{1}{2}mv^2 = Vq$$

VI Modern physics

Worked example
Calculate the voltage through which a proton, of mass 1.66×10^{-27} kg, must be accelerated to reach a speed of 8.0×10^6 m s^{-1}.

Answer
$½mv^2 = Vq$

Rearrange the equation:
$V = ½mv^2/q = 0.5 \times 1.66 \times 10^{-27} \times (8.0 \times 10^6)^2 / 1.6 \times 10^{-19} = 3.3 \times 10^5$ V

A useful unit
In the example above the proton was accelerated through a potential difference of 3.3×10^5 V. To calculate its energy we use the formula:

$E = qV$

energy of this proton = $1.6 \times 10^{-19} \times 3.3 \times 10^5 = 5.3 \times 10^{-14}$ J

1.6×10^{-19} C is the electronic charge, so it may be said that the energy of the proton (or other particle) is 3.3×10^5 electronvolts (eV).

Thus the electronvolt is a unit of energy. It is the energy that an electron gains when accelerated through a potential difference of 1 volt; it is equal to 1.6×10^{-19} J.

Worked example 1
Deduce the energy a proton gains when it is accelerated through a potential difference of 2.0 MV. Express your answer in electronvolts and in joules.

Answer
2.0 MV = 2.0×10^6 V

charge on the proton is +e, so it gains 2.0×10^6 eV of energy

2.0×10^6 eV = $2.0 \times 10^6 \times 1.6 \times 10^{-19} = 3.2 \times 10^{-13}$ J

Worked example 2
Deduce the energy an alpha particle gains when it is accelerated through a potential difference of 2.0 MV. Express your answer in electronvolts.

Answer
charge on an alpha particle is +2e, so

energy gained = $2 \times 2.0 \times 10^6 = 4.0 \times 10^6$ eV

Uniform field at right angles to the motion of a charged particle

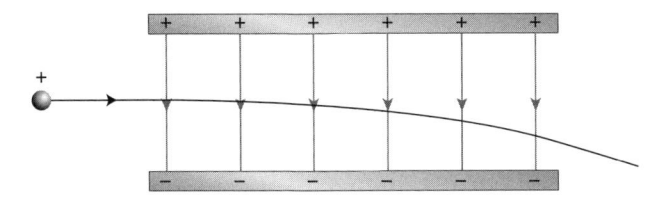

The force on the particle in the example above is vertically downwards. The path that the particle takes is similar to that of a body thrown horizontally in a uniform gravitational field (page 137). There is no change to the horizontal component of the velocity but there is a constant acceleration perpendicular to the velocity. This produces the typical parabolic path.

Worked example
The diagram below shows the path of an electron travelling through a uniform electric field.

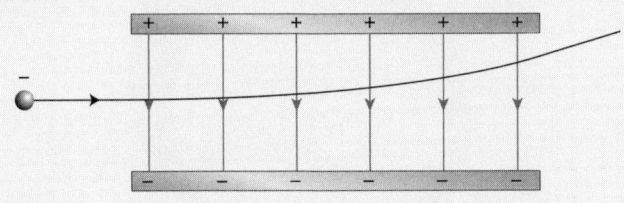

On a copy of the diagram, sketch:
(a) the path that a positron moving at the same speed would take; label this **A**
(b) the path that a proton travelling at the same speed would take; label this **B**

(A positron is identical to an electron but with a positive charge.)

Answer

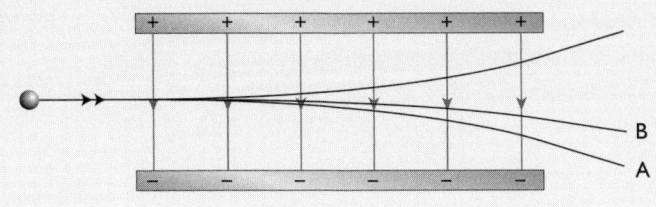

Note
The electron, being negative, is deflected in the opposite direction to the proton described earlier. In this example, the positron and the proton are deflected in the

same direction but because the proton is much more massive than the positron its deflection is much less. In reality, the deflection would be barely visible.

Magnetic fields

Measurement of e/m

We saw on page 187 that a particle of mass m, carrying charge q moving with velocity v perpendicular to a uniform magnetic field experiences a force $F = Bqv$, and travels in a circular path.

This force acts as the centripetal force, so:

$Bqv = mv^2/r$

Hence:

$q/m = v/Br$

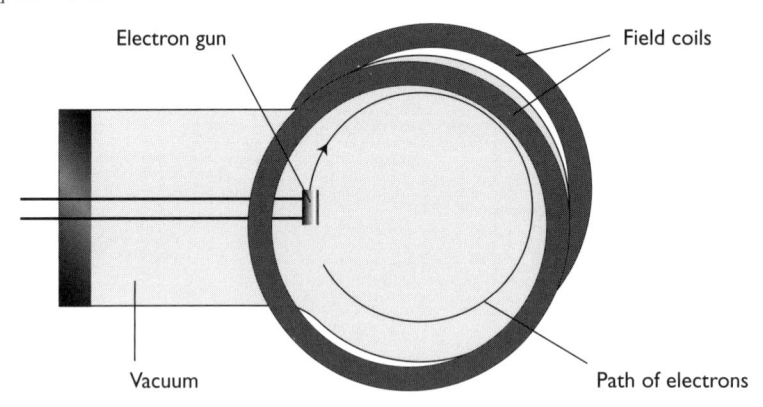

Apparatus for measurement of e/m

Apparatus for measuring e/m is shown in the diagram above.

The electron gun fires electrons vertically upwards. The field coils produce a uniform electric field, which causes the electrons to travel in a circular path. There is a trace of gas in the tube. When the electrons collide with the gas atoms they can cause ionisation (or excitation) of these atoms. When the atoms drop back into their ground state (see page 210) they emit a pulse of light. In this way the path of the electrons can be observed. The energy of the electrons is calculated from the accelerating potential, and the diameter of their circular path is measured with a ruler.

Worked example

Electrons travelling at a velocity of $4.0 \times 10^6 \, m\,s^{-1}$ enter a uniform magnetic field of 0.60 mT at right angles to the field. The electrons then travel in a circle of diameter 7.6 cm.

Calculate the value of e/m for the electron.

Answer
radius of the circle = $7.6 \times 10^{-2}/2 = 3.8 \times 10^{-2}$ m

$$\frac{e}{m} = \frac{V}{Br} = \frac{4.0 \times 10^6}{0.6 \times 10^{-3} \times 3.8 \times 10^{-2}} = 1.75 \times 10^{11} \, C\,kg^{-1}$$

The charge on an electron is 1.6×10^{-19} C; the mass of an electron can now be calculated:

$e/m = 1.75 \times 10^{11}$

$m = e/1.75 \times 10^{11} = 1.6 \times 10^{-19}/1.75 \times 10^{11} = 9.1 \times 10^{-31}$ kg

Velocity selectors

For more sophisticated measurements of q/m it is important to have particles of a more precise speed. One method of achieving this is to use crossed electric and magnetic fields.

Collimator to produce
a parallel beam of electrons

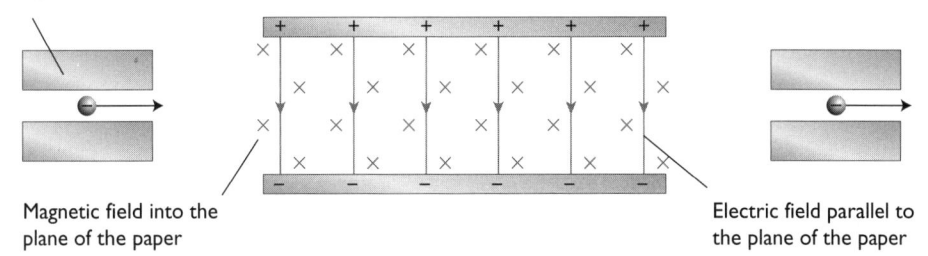

Magnetic field into the
plane of the paper

Electric field parallel to
the plane of the paper

The force from the electric field on the electrons is vertically upwards.

$F_E = eE$

The force due to the magnetic field is vertically downwards.

$F_B = Bev$

The fields are adjusted so that the magnetic force exactly balances the electric force for those electrons that have the required velocity, so that they, and only they, go through the second collimating slits. Electrons with a slightly higher than required velocity experience a slightly larger magnetic force than electric force and are deflected downwards. Therefore, they miss the second slit. Electrons that are slightly slower than required are deflected upwards and also miss the slit.

For the selected velocity:

$F_E = F_B$

Therefore: $eE = Bev$

Thus, **$v = E/B$**

This equation has been worked out for electrons. However, neither the charge nor the mass of the particle feature in the equation, so it is valid for any charged particle.

> **Worked example**
> In order to find the velocity of alpha particles from a radioactive source, a narrow beam of the particles is fired into crossed electric and magnetic fields. It is found that the particles travel straight through the fields when the electric field strength is $8.6 \times 10^3 \, V\,m^{-1}$ and the magnetic field strength is 0.48 mT.
>
> What is the velocity of the alpha particles?
>
> **Answer**
> $v = E/B = 8.6 \times 10^3 / 0.48 \times 10^{-3}$
>
> $= 1.8 \times 10^7 \, m\,s^{-1}$

Quantum physics

Waves or particles?

In the section on waves, you learnt that light shows the properties of waves — diffraction, interference and polarisation. In this section we investigate properties of light which suggest that it also behaves like particles. You will also learn that electrons show wave properties.

Photoelectric effect

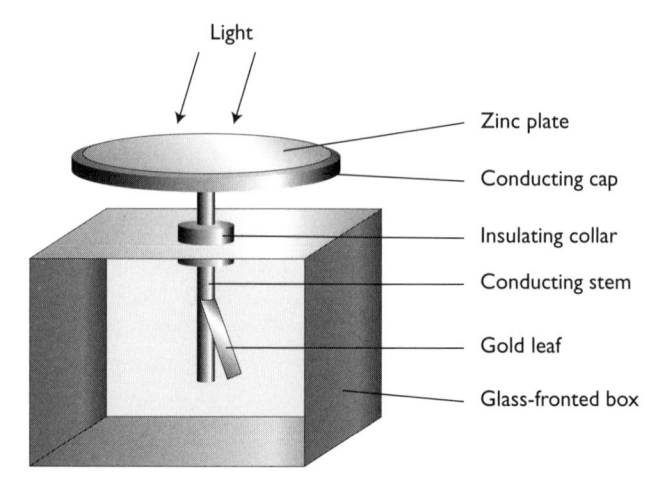

Gold-leaf electroscope being used to demonstrate the photoelectric effect

The electroscope is charged negatively. When visible light is shone onto the zinc plate the electroscope the remains charged no matter how bright the light. When ultraviolet light is shone onto the plate, it steadily discharges; the brighter the light the faster it discharges. Ultraviolet light has enough energy to lift electrons out of the plate and for them to leak away into the atmosphere; visible light does not have sufficient energy.

This cannot be explained in terms of a wave model. If light is transferred by waves, eventually, whatever the frequency, enough energy would arrive and electrons would escape from the metal surface. In practice, provided the radiation has a high enough frequency, electrons are emitted instantaneously. Electromagnetic radiation arrives in packets of energy — the higher the frequency, the larger the packet. These packets of energy, or **quanta**, are called **photons**. The emission of photoelectrons occurs when a single photon interacts with an electron in the metal — hence the instantaneous emission of the photoelectron. The packets of energy for visible light are too small to eject electrons from zinc; the ultraviolet packets are large enough to do so.

More on the photoelectric effect

More detailed experiments measuring the maximum energy of electrons emitted in the photoelectric effect (photoelectrons) give evidence for the relationship between photon energy and frequency. They also show that the intensity of the electromagnetic radiation does not affect the maximum energy with which the electrons are emitted, just the rate at which they are emitted.

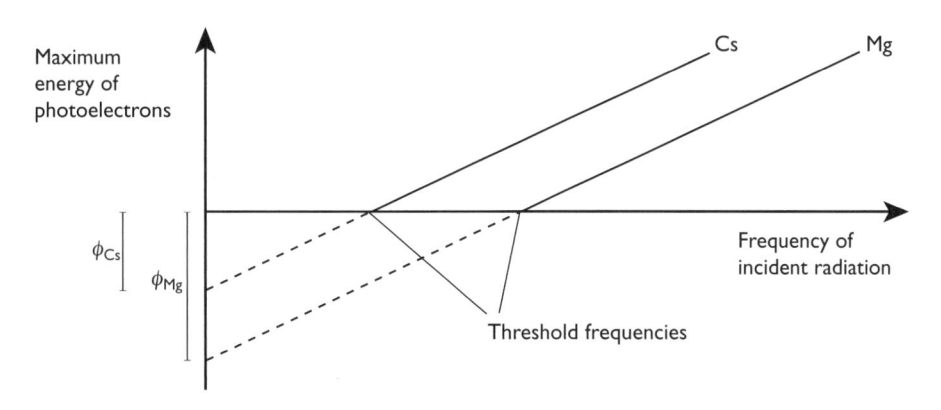

Graph of maximum energy of photoelectrons against frequency of incident radiation

The diagram above gives a great deal of information:
- The graphs are straight-line graphs with the same gradient, whatever metals are used.
- Each metal has a particular minimum frequency at which electrons are emitted. This frequency is called the **threshold frequency**.
- The minimum energy required to eject an electron from each metal is different. This energy is indicated by the symbol ϕ in the diagram. This is called the **work function energy** of the metal. It is important to recognise that the work function

energy is the *minimum* energy, or minimum work, needed to remove an electron from the surface of the metal.

The general equation for the graphs on page 207 is of the form:

$y = mx + c$

In this case:

$E = hf - \phi$

The gradient is the same for all metals so h is universal. It is known as the **Planck constant** = 6.63×10^{-34} Js.

If you study the graph you will see that the quantity hf is the energy of the incoming photons.

So photon energy **$E = hf$**

Notice how this equation, sometimes known as the Einstein–Planck equation, combines the wave nature (frequency) and the particle nature (energy of a photon) in the same equation. The equation also shows that the maximum energy of a photoelectron is independent of the intensity of the radiation and depends solely on the frequency of the radiation. Greater intensity means a greater rate of arrival of photons, but the energy of each photon is still the same.

Worked example
The wavelength of red light is approximately 7×10^{-7} m. Calculate the energy of a red light photon.

Answer
$E = hf = hc/\lambda$

$= 6.63 \times 10^{-34} \times 3 \times 10^{8} / 7 \times 10^{-7} = 2.8 \times 10^{-19}$ J

Hint

The alternative way of expressing the energy of a photon ($E = hc/\lambda$) is derived from $E = hf$ and $f\lambda = c$. It is a useful expression to remember.

When photoelectrons are ejected from the surface of a metal, the 'spare' energy of the photon (the energy that is not used in doing work to lift the electron out of the metal) is given to the electron as kinetic energy. When light is incident on the metal surface, electrons are emitted with a range of kinetic energies, depending on how 'close' they were to the surface when the photons were incident upon them. The maximum kinetic energy is when the minimum work is done in lifting the electron from the surface, consequently:

$$hf_0 = \phi + \frac{1}{2}mv_{max}^2$$

It follows that $hf_0 = \phi$, where f_0 is the threshold frequency of the radiation.

Worked example
The work function energy of caesium is 2.1 eV.

(a) Calculate the threshold frequency for this metal.

(b) State in what range of the electromagnetic spectrum this radiation occurs.

(c) Radiation of frequency 9.0×10^{14} Hz falls on a caesium plate. Calculate the maximum speed at which a photoelectron can be emitted.

(mass of an electron = 9.1×10^{-31} kg)
(charge on an electron = 1.6×10^{-19} C)

Answer
(a) $\phi = 2.1 \times 1.6 \times 10^{-19} = 3.36 \times 10^{-19}$ J

$E = hf$

$f = E/h = 3.36 \times 10^{-19}/6.63 \times 10^{-34} = 5.1 \times 10^{14}$ Hz

(b) It is in the visible spectrum, in the yellow region.

(c) Energy of the photon:

$E = hf = 6.63 \times 10^{-34} \times 9.0 \times 10^{14} = 5.97 \times 10^{-19}$ J

$hf = \phi + E_k$

$E_k = hf - \phi = 5.97 \times 10^{-19} - 3.36 \times 10^{-19} = 2.61 \times 10^{-19}$ J

$E_k = \frac{1}{2}mv^2$

$v = \sqrt{2E_k/m}$ $(2 \times 2.61 \times 10^{-19}/(9.1 \times 10^{-31})) = 7.6 \times 10^5$ m s^{-1}

Spectra
Line emission spectra
We looked at the spectrum of hot bodies in the AS course, the visible spectrum being a continuous band of light with one colour merging gradually into the next. Such a spectrum is called a continuous spectrum. If a large potential difference is put across low pressure gases they emit a quite different spectrum, consisting of a series of bright lines on a dark background. This type of spectrum is called a **line emission spectrum**.

VI Modern physics

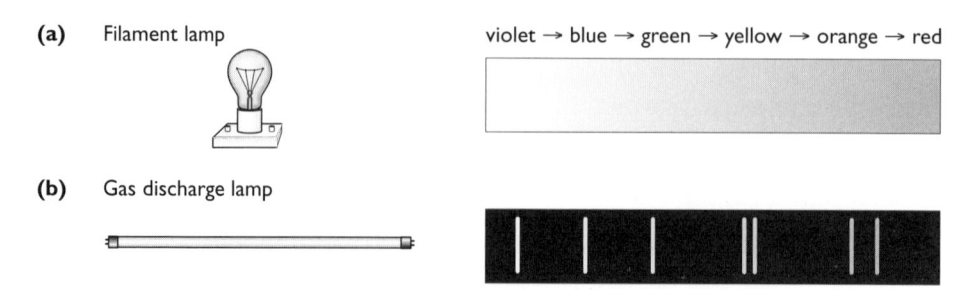

(a) Filament lamp

violet → blue → green → yellow → orange → red

(b) Gas discharge lamp

(a) Continuous spectrum from a hot filament lamp
(b) Line emission spectrum from a gas-discharge tube

The precise lines visible depend on the gas in the discharge tube; each gas has its own unique set of lines. This is useful for identifying the gases present, not only in Earth-based systems but in stars as well.

Each line has a particular frequency. Hence each photon from a particular line has the same energy.

Atoms in gases are far apart from each other and have little influence on each other. The electrons in each atom can only exist in fixed energy states. Exciting the gas means that electrons are given energy to move from the lowest energy state (the ground state) to a higher energy state. They will remain in that state for a time before dropping back to a lower state. When they do so they emit a photon.

The diagram below shows the energy levels for a hydrogen atom and an electron falling from level $n = 3$ to $n = 1$.

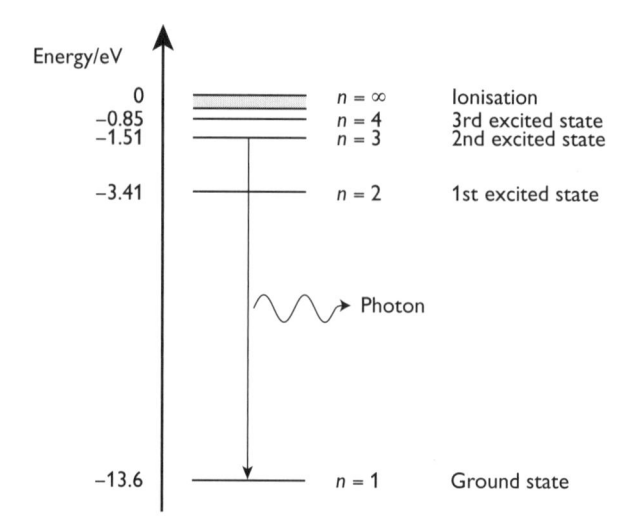

Energy/eV

0	$n = \infty$	Ionisation
−0.85	$n = 4$	3rd excited state
−1.51	$n = 3$	2nd excited state
−3.41	$n = 2$	1st excited state
		Photon
−13.6	$n = 1$	Ground state

Note

Notice that when the electron is totally removed from the nucleus the ionisation level is zero and the other energies are all negative. Compare this with the potential energies near a charged sphere.

In general, the frequency of the emitted photon when an electron drops from a level E_1 to a level E_2 can be calculated from the equation:

$hf = E_1 - E_2$

Worked example
Calculate the frequency of the photon emitted when the hydrogen atom falls from the second excited state to the ground state.

Answer

$E = hf$

E = the difference in the energy levels = $-1.51 - (-13.6) \approx 12.1$ eV

$E = 12.1 \times 1.6 \times 10^{-19} = 19.4 \times 10^{-19}$ J

$f = E/h = 19.4 \times 10^{-19}/(6.63 \times 10^{-34}) = 2.9 \times 10^{15}$ Hz

This is in the ultraviolet part of the spectrum.

Line emission spectra give strong evidence for the existence of discrete energy levels in atoms. The photons emitted are of a definite set of frequencies and therefore of a definite set of energies. Each specific energy photon corresponds to the same fall in energy of an orbital electron as it drops from one discrete energy level to a second, lower, discrete energy level.

Line absorption spectra
When white light from a continuous spectrum is shone through a gas or vapour the spectrum observed is similar to a continuous spectrum, except that it is crossed by a series of dark lines.

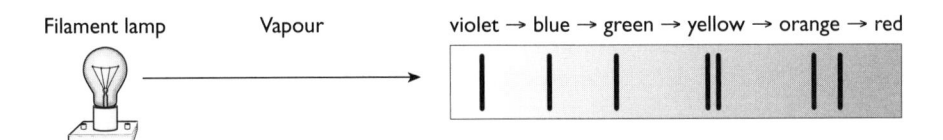

A line absorption spectrum

This type of spectrum is called a **line absorption spectrum.** White light consists of all colours of the spectrum, which is a whole range of different frequencies and therefore photon energies. As the light goes through the gas/vapour the photons of energy exactly equal to the differences between energy levels are absorbed, as shown in the next diagram.

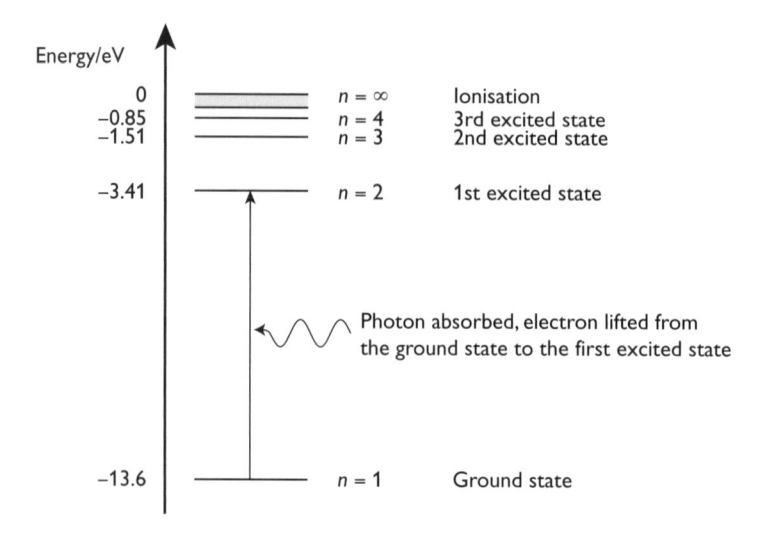

```
Energy/eV
    0        ────────    n = ∞      Ionisation
 −0.85       ────────    n = 4      3rd excited state
 −1.51                   n = 3      2nd excited state

 −3.41       ────────    n = 2      1st excited state

                                    Photon absorbed, electron lifted from
                                    the ground state to the first excited state

 −13.6       ────────    n = 1      Ground state
```

The light is then re-emitted by the newly excited atom. However, this secondary photon can be emitted in any direction, so the energy of this frequency radiated towards the observer is very small, hence dark lines are observed in the spectrum.

Worked example

The line absorption spectrum from a star is studied. A dark line is observed at a wavelength of 6.54×10^{-7} m. Calculate the difference in the two energy levels which produces this line.

Answer

$E = hf = hc/\lambda = 6.63 \times 10^{-34} \times 3.0 \times 10^8 / (6.54 \times 10^{-7}) = 3.04 \times 10^{-19}$ J

Note

If the answer to the above worked example is converted to electronvolts, it becomes 1.90 eV. This is the difference between the first and second excited levels in the hydrogen atom. This reaction is quite likely — the outer atmosphere of the star, although cooler than the core, is still at a high temperature, so a lot of atoms will be in the first, and other, excited states. This provides evidence that the outer atmosphere of the star contains hydrogen, although more lines would need to be observed for this to be confirmed.

Electrons

'If light can behave like waves and like particles, can electrons behave like waves?'

That was the thought, in the 1920s, of a postgraduate student, **Louis de Broglie**, who first proposed the idea of 'matter waves'. If electrons do travel through space as waves, then they should show diffraction effects.

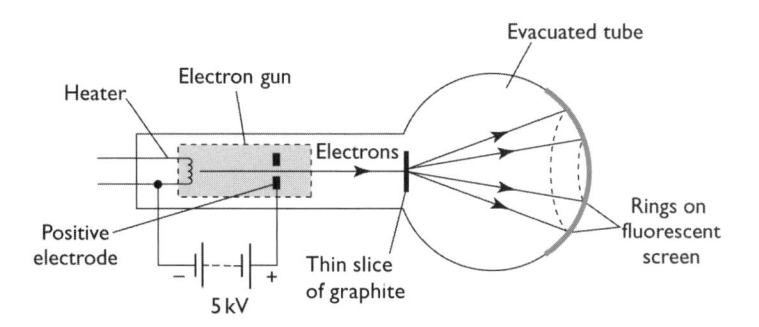

Apparatus to demonstrate electron diffraction

In this experiment, electrons are emitted from the hot cathode and accelerated towards the thin slice of graphite. The graphite causes diffraction and the maxima are seen as bright rings on the fluorescent screen. The diameter of the rings is a measure of the angle at which the maxima are formed. The diameters are dependent on the speed to which the electrons are accelerated. The greater the speed the smaller the diameter, and hence the smaller the diffraction angle. From this information it can be concluded that:
- electrons travel like waves
- the wavelength of those waves is similar to the spacing of the atoms in graphite (otherwise diffraction would not be observed)
- the wavelength of the waves decreases with increasing speed of the electrons

The de Broglie equation

It was proposed by de Broglie that the wavelength associated with electrons of mass m travelling at a velocity v could be given from the formula:

$\lambda = h/mv$

You will recall that the quantity mv is the momentum p. The equation can be rewritten as:

$\lambda = h/p$

Hint

It should be noted that the latter form of the equation is the more significant. This is because all matter has an associated wave function and it is the momentum, rather than the speed, that is the determining factor of the wavelength.

Worked example

Electrons accelerated through a potential difference of 4.0 kV are incident on a thin slice of graphite that has planes of atoms 3.0×10^{-10} m apart.

Show that the electrons would be suitable for investigating the structure of graphite. (mass of an electron is 9.1×10^{-31} kg)

VI Modern physics

> **Answer**
> energy of the electrons = 4.0 keV = $4.0 \times 10^3 \times 1.6 \times 10^{-19} = 6.4 \times 10^{-16}$ J
>
> $E_k = \frac{1}{2}mv^2$
>
> $v = \sqrt{2E_k/m} = \sqrt{2 \times 6.4 \times 10^{-16}/(9.1 \times 10^{-31})} = 3.8 \times 10^7$ m s^{-1}
>
> $\lambda = h/mv = 6.63 \times 10^{-34}/(9.1 \times 10^{-31} \times 3.8 \times 10^7)) = 1.9 \times 10^{-11}$ m
>
> The waves are of a similar order of magnitude as the atomic layers of graphite and are therefore suitable.

Other matter waves

It is not just electrons that have an associated wave function, all matter does. Neutron diffraction is an important tool in the investigation of crystal structures because they are uncharged. From the de Broglie equation, you can see that, for the same speed, neutrons will have a much shorter wavelength than electrons because their mass is much larger. Consequently, slow neutrons are used when investigating at the atomic level.

What about people-sized waves? Consider a golf ball of approximate mass 50 g being putted across a green at 3 m s^{-1}. Its wavelength can be calculated:

$\lambda = h/p = 6.63 \times 10^{-34}/0.050 \times 3 = 4.4 \times 10^{-33}$ m

This is not even 1 trillionth the diameter of an atomic nucleus! Consequently, we do not observe the wave function associated with everyday-sized objects (and it does not explain why the author's short putts have a tendency not to go in the hole!)

Nuclear physics

Mass and energy

In the AS course the idea of mass–energy was introduced — the greater the energy of an object the greater is its mass. These two quantities are linked by Einstein's mass–energy equation:

$E = mc^2$

This equation quantifies the extra mass a body gains when its energy is increased.

> **Worked example**
> A proton in a particle accelerator is accelerated through 4.5 GV. Calculate the increase in mass of the proton.

> **Answer**
> energy gained by the proton = 4.5 GeV
>
> convert this into joules:
>
> $E = 4.5 \times 10^9 \times 1.6 \times 10^{-19} = 7.2 \times 10^{-10}$ J
>
> $E = mc^2$
>
> Therefore:
> $m = E/c^2 = 7.2 \times 10^{-10}/(3.0 \times 10^8)^2 = 8.0 \times 10^{-27}$ kg

Note

This is an amazing result. The rest mass of a proton is 1.66×10^{-27} kg. The increase in mass is almost five times this, giving a total mass of almost six times the rest mass.

Binding energy and mass defect

Just as with the energy levels in the outer atom and with the electrical energy of a negative particle near a positive charge, the field inside a nucleus can be considered attractive (but be careful — the field is *not* produced by electric charges). The zero of energy is once more taken as infinity and, therefore, the particles in the nucleus have negative energy. The energy required to separate a nucleus into its constituent protons and neutrons is called the **binding energy**. Each nuclide has a different binding energy; the binding energy per nucleon in the nuclide is a measure of its stability. A binding energy curve showing the general trend and specific important nuclides is shown below.

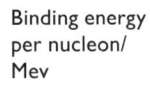

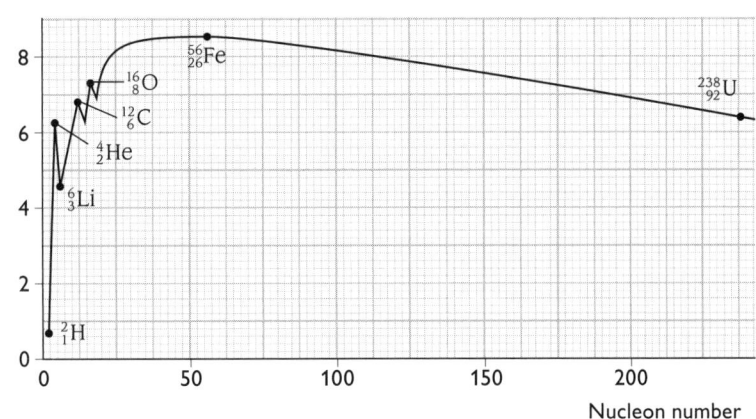

Binding energy curve

In particular, note the high binding energies for $^4_2$He, $^{12}_6$C and $^{16}_8$O. The highest binding energy per nucleon, and therefore the most stable nuclide, is $^{56}_{26}$Fe.

Binding energies are very large and hence there is a measurable difference in the mass of a proton that is bound in a nucleus and that of a free proton at rest. The shape of the curve for the 'missing mass' per nucleon, known as the **mass defect** per nucleon, is exactly the same as that for binding energy.

Worked example
A carbon-12 atom consists of 6 protons, 6 neutrons and 6 electrons. The unified mass unit (u) is defined as 1/12 the mass of the carbon-12 atom.

Calculate:
(a) the mass defect in kilograms
(b) the binding energy
(c) the binding energy per nucleon

(mass of a proton = 1.007 276 u, mass of a neutron = 1.008 665 u, mass of an electron = 0.000 548 u, 1 u = 1.66 × 10^{-27} kg)

Answer
(a) mass of 6 protons + 6 neutrons + 6 electrons = 6 × (1.007 276 + 1.008 665 + 0.000 548) u = 12.098 934 u

mass defect = 12.098 934 − 12 = 0.098 934 u = 0.098 934 × 1.66 × 10^{-27} kg = 1.64 × 10^{-28} kg

(b) binding energy, $E = mc^2$ = 1.64 × 10^{-28} × (3.0 × 10^8)^2 = 1.48 × 10^{-11} J

(c) binding energy per nucleon = 1.48 × 10^{-11}/12 = 1.23 × 10^{-12} J

Nuclear fission

Fission is the splitting of a nucleus into two roughly equal-sized halves with the emission of two or three neutrons. If you look at the binding energy curve on page 215, you will see that the nuclides with nucleon numbers between about 50 and 150 have significantly more binding energy per nucleon than the largest nuclides with nucleon numbers greater than 200. A few of these larger nuclides are liable to fission. Fission happens rarely in nature. However, physicists can induce fission by allowing large, more stable nuclides to capture a neutron to form an unstable nuclide. For example, a uranium-235 nucleus, which is found in nature, can capture a slow moving nucleus to form a uranium-236 nucleus.

$$^{235}_{92}U + ^{1}_{1}n \rightarrow ^{236}_{92}U$$

This nucleus is unstable and will undergo fission:

$$^{236}_{92}U \rightarrow ^{146}_{57}La + ^{87}_{35}Br + 3^{1}_{0}n$$

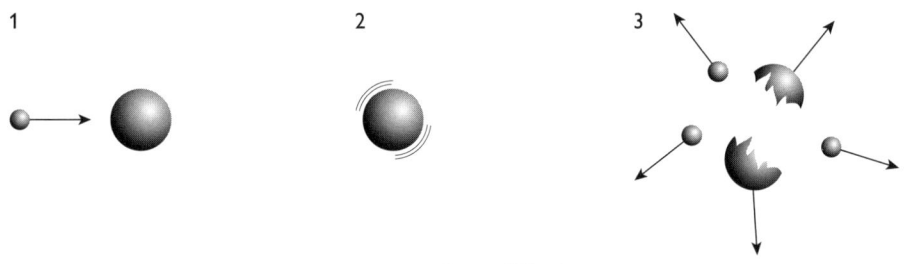

A cartoon view of fission

In the diagram above:
- Step 1 — a neutron trundles towards a U-235 nucleus
- Step 2 — the U-235 nucleus absorbs the neutron to form an unstable U-236 nucleus
- Step 3 — the U-236 nucleus splits into two roughly equal halves (the fission fragments) which fly apart; three neutrons are released, which also fly away at high speeds

Most of the energy in fission is carried away by the fission fragments as kinetic energy, although some is carried away by the neutrons. In addition, gamma rays are formed. Some are formed almost immediately and some are formed later as the nucleons in the fission fragments rearrange themselves into a lower, more stable energy state.

Fission is used in all working nuclear power stations. If the fissionable nuclide being used is uranium, the neutrons released are slowed down, so that they cause new fissions to keep the process going. For power generation, each fission must, on average, produce one new fission to keep the reaction going at a constant rate.

The earliest nuclear weapons used fission. The principle is the same but in this case each fission must induce more than one new fission on average, so that the reaction rate rapidly gets faster and faster causing an explosion.

Nuclear fusion

Nuclear fusion can be viewed as the opposite of fission. Two small nuclei move towards each other at great speeds, overcome the mutual electrostatic repulsion and merge to form a larger nucleus. If you look at the binding energy curve on page 215 again, you will see that the binding energy per nucleon of deuterium ($^2_1$H) is much less than that of helium ($^4_2$He). Thus two deuterium nuclei could fuse to form a helium nucleus with the release of energy. In practice, the fusion of tritium and deuterium is more common:

$$^2_1H + ^3_1H \rightarrow ^4_2He + ^1_0n$$

Fusion releases much more energy per nucleon involved than fission. The difficulty in using fusion commercially is that the pressure and temperature of the fusing mixture are extremely high — so high that all the electrons are stripped from the nuclei and the mixture becomes a sea of positive and negative particles called a plasma. A physical container cannot be used to hold the plasma — it would immediately vaporise (not to mention cool the plasma). Extremely strong magnetic fields are used to contain the plasma. Even though scientists have not yet solved the problems of controlling nuclear fusion, we rely on it because it is the process that fuels the Sun.

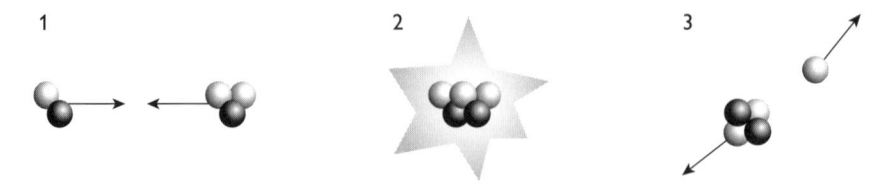

A cartoon view of fusion

In the diagram:
- Step 1 — a deuterium nucleus and a tritium nucleus move towards each other at high speed
- Step 2 — they collide and merge
- Step 3 — a helium nucleus and a neutron are formed and fly apart at high speed

Radioactive decay

You met the idea of radioactive decay in your AS work. This section develops those ideas and introduces the mathematics of radioactive decay.

You already know that radioactive decay is random. We cannot tell when a nucleus will decay, only that there is a fixed chance of it decaying in a given time interval. Thus, if there are many nuclei we can say:

$A = \lambda N$

where A is the **activity** (the number of nuclei that decay per unit time), N is the total number of nuclei in the sample and λ is a constant, known as the decay constant. The decay constant is the probability per unit time that a nucleus will decay. It is measured in s^{-1}, min^{-1}, yr^{-1} etc.

The activity is measured in a unit called the becquerel (Bq).

One becquerel is an activity of one decay per second.

If we measure the activity of an isotope with a relatively short half-life (say 1 minute) then we can plot a graph similar to the one below.

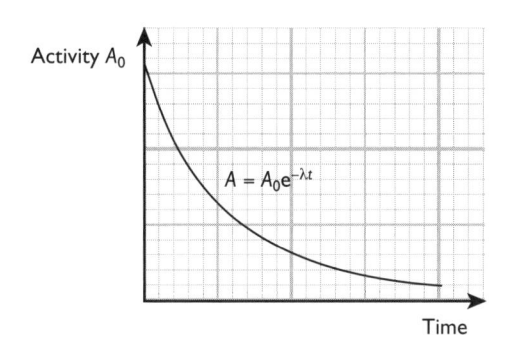

Exponential decay of a radioactive isotope

If you study this graph, you will find that the activity falls by equal proportions in successive time intervals. The number of atoms decaying in a fixed period decreases because the number of atoms remaining is decreasing. This type of decay is called exponential decay and the equation that describes it is of the form:

$$x = x_0 e^{-\lambda t}$$

In the example given in the diagram above, the y-axis is labelled activity A, but because the activity is directly proportional to the number of nuclei present (N) it could equally be N. The count rate is directly proportional to the activity, so y could also apply to this.

Half-life and radioactive decay

Look at the graph above again. You have already observed that the activity decreases by equal proportions in equal time intervals. Now look at how long it takes to fall to half the original activity. How long does it take to fall to half this reading (one-quarter of the original)? How long does it take to fall to half of this? You should find that each time interval is the same. This quantity is called the **half-life ($t_{1/2}$)**.

The half-life of a radioactive isotope is the time taken for half of the nuclei of that isotope in any sample to decay.

Investigation of radioactive decay

When investigating the decay of a radioactive isotope it is not possible to measure directly either the number of atoms of the isotope remaining in the sample, or the activity of the sample. The detector detects only a small proportion of the radiation given off by the sample. Radiation is given off in all directions and most of it misses the detector. Even radiation that enters the detector may pass straight through it without being detected. What is measured called the **received count rate**. The background count rate should be subtracted from the received count rate to give the corrected count rate.

Half-life and the decay constant

If the decay equation is applied to the half-life it becomes:

$$\tfrac{1}{2} N_0 = N_0 e^{-\lambda t_{1/2}}$$

Cancelling the N_0 gives:

$$\frac{1}{2} = e^{-\lambda t_{1/2}}$$

Taking logarithms of both sides:
 $\ln(½) = -\lambda t_{1/2}$
 $\ln(½) = -\ln 2$

Therefore, $\ln 2 = \lambda t_{1/2}$

and $t_{1/2} = \ln 2/\lambda$

Remember that $\ln 2$ is the natural logarithm of 2 and is approximately equal to 0.693.

Worked example
The proportion of the carbon-14 isotope found in former living material can be used to date the material.

The half-life of carbon-14 is 5730 years. A certain sample has 76.4% of the proportion of this isotope compared with living tissue.

(a) Calculate the decay constant for this isotope of carbon.

(b) Calculate the age of the material.

Answer

(a) $\lambda = \ln 2 / t_{1/2} = 0.693/5730 = 1.21 \times 10^{-4}$ years^{-1}

(b) $N = N_0 e^{-\lambda t}$

$\ln(N/N_0) = -\lambda t$

$t = -\ln(76.4/100)/(1.21 \times 10^{-4}) = 2200$ years

Radioactivity and binding energy

Whenever there is a spontaneous nuclear decay, the nucleus decays into a more stable/lower energy nucleus. The total binding energy after the decay is less than before the decay. The 'lost' energy is (mainly) carried away by the particles/gamma rays that are emitted.

Worked example
The equation shows the decay of the isotope Ra-226:

$^{226}_{88}\text{Ra} \rightarrow\ ^{222}_{86}\text{Rn} + ^{4}_{2}\alpha$

Calculate the energy of the alpha particle that is released in this reaction. State any assumptions you make.

(mass of the Ra-226 nucleus = 225.977 u, mass of the Rn-222 nucleus = 221.970 u, mass of the alpha particle = 4.003 u)

($1 u = 1.66 \times 10^{-27}$ kg)

Answer
mass defect = $225.977 - 221.970 - 4.003 = 0.004 u = 0.004 \times 1.66 \times 10^{-27}$
$= 6.64 \times 10^{-30}$ kg

$E = mc^2 = 6.64 \times 10^{-30} \times (3 \times 10^8)^2 = 6.0 \times 10^{-13}$ J

Note

In the worked example above, it is assumed that the alpha particle carries away all the energy. In practice, the nucleus would recoil. This should remind you of the work on momentum and collisions. In addition, in many alpha decays a gamma ray is also emitted.

Gathering and communicating information

Direct sensing

All electronic sensors can be considered to have three parts:

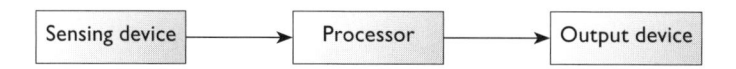

Sensing device → Processor → Output device

The sensing device is an active device; it might be a light-dependent resistor or a thermistor. The processor can be as simple as a potential divider or it can be more complex, such as a operational amplifier. Its job is to convert the change in the relevant physical property into a change in voltage. In the case of an operational amplifier it also amplifies the voltage. The output device could be a measuring device such as a voltmeter, or an indicator such as an LED.

Sensing devices
Light-dependent resistor (LDR)
The resistance of a LDR decreases with increasing light levels. Typical values range from 100 Ω in bright sunlight to in excess of 1 MΩ in darkness.

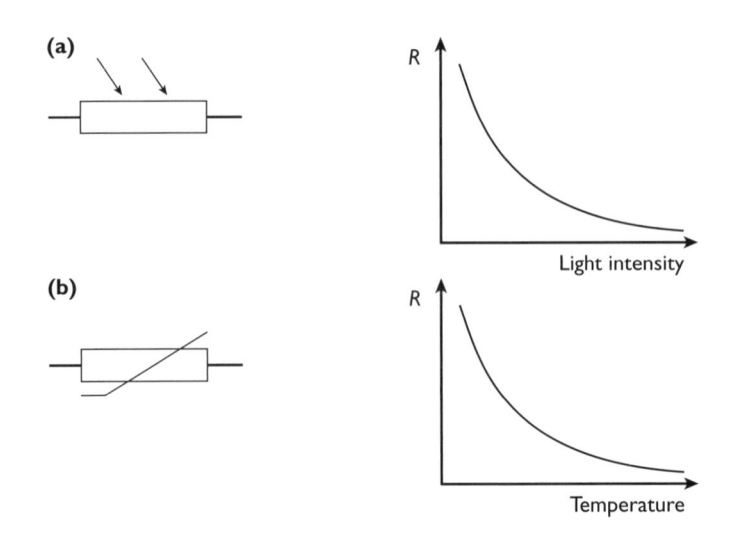

Circuit symbol and characteristic of (a) an LDR and (b) a thermistor

Thermistor

Although there are different types of thermistor, you only need to know about **negative temperature coefficient** thermistors. The resistance of a negative temperature coefficient thermistor decreases with increasing temperature.

A wide range of thermistors is available and an engineer will choose a suitable one for a particular job. However, typical resistance values might vary from 1 kΩ at room temperature falling to around 10 Ω at 100°C.

Strain gauge

The diagram shows the structure of a **strain gauge**. It consists of a length of wire embedded in and running up and down a plastic base. When the plastic bends, the wire stretches. This increases its length and reduces its cross-sectional area, thereby increasing its resistance. You should recall these ideas from the work on resistivity.

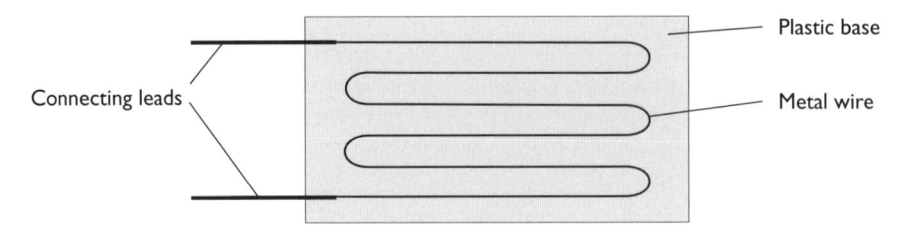

A strain gauge

The change in the resistance ΔR is proportional to the change in length ΔL of the wire, and for small changes:

$\Delta R/R = 2\Delta L/L$

Note that this formula assumes that the length increases and that the cross-sectional area decreases. If it is assumed that the cross-sectional area is unchanged, the relationship is $\Delta R/R = \Delta L/L$. In the examination it is important you make clear your assumption about the area.

Worked example
The diagram shows a strain gauge in a potential divider circuit.

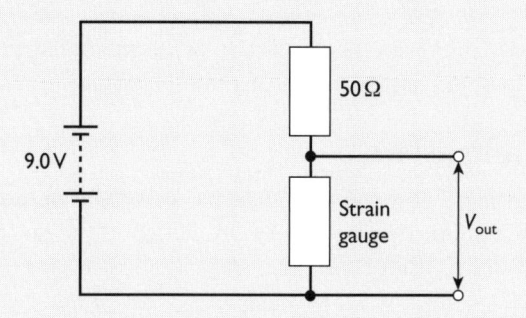

The initial resistance of the strain gauge is 150 Ω. When under strain, the length of the wire increases by 0.80%.

Calculate:
(a) the initial voltage output from the potential divider
(b) the voltage output when the gauge is under strain

Answer
(a) $V_{out} = V_{in} \times R_2/(R_1 + R_2) = 9 \times 150/(150 + 50) = 6.75$ V

(b) Assuming the stretching of the wire causes a corresponding reduction in the cross-sectional area:

$\Delta R = 2R\Delta L/L = 2 \times 150 \times 0.80/100 = 2.4$ Ω

$R_{new} = 150 + 2.4 = 152.4$ Ω

$V_{out} = V_{in} \times R_2/(R_1 + R_2) = 9 \times 152.4/(152.4 + 50) = 6.78$ V

Note
The change in output is small but measurable, although you may need to use an operational amplifier, which is discussed in the next section.

Piezoelectric effect
Certain materials, for example quartz, when they are under either compressive or tensile stress generate an e.m.f. across their crystal faces. Under compression, the e.m.f. is generated in one direction; when under tensile stress the e.m.f. is in the opposite direction.

This property is used in the piezoelectric microphone. A sound wave is a pressure wave. It can cause a piezoelectric crystal to compress and stretch in a pattern similar to that of the incoming wave. This produces a varying e.m.f. across the crystal which can be amplified as necessary.

The usefulness of the piezoelectric effect does not end here; not only does applying stress produce a voltage, the reverse is true. If a potential difference is applied across such a crystal it either becomes compressed or expands depending on the direction of the e.m.f.

This second property is used in the generation of ultrasound for scanning purposes. This is discussed in detail in the section on remote sensing.

Operational amplifiers

Amplifiers produce more power output than input and the **operational amplifier** (**op-amp**) is no exception. This is not a contradiction of the law of conservation of energy — energy is fed in from elsewhere.

The circuit diagram symbol for an operational amplifier is shown below:

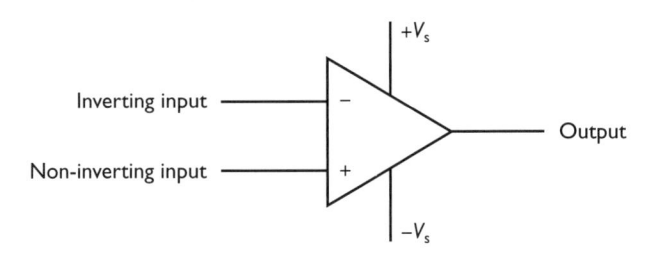

The connections to the power supply ($+V_s$ and $-V_s$) are sometimes not included in diagrams. However, in practice they are necessary.

There is also (not shown on the diagram) an earth or zero line. This is important because all voltages are measured relative to this. It follows that $+V_s$ and $-V_s$ have equal magnitude.

The op-amp has two inputs, the **inverting input** (shown on the diagram by a − sign) and a **non-inverting input** (shown by a + sign). The output from a signal applied to the inverting input is 180° out of phase with the input signal. The output from the non-inverting output is in phase with the input signal.

The **gain of an amplifier A** is defined by the equation:

A = output voltage/input voltage

The ideal op-amp
The ideal op-amp has the following properties:
- infinite open-loop gain
- infinite input impedance
- zero output impedance
- infinite bandwidth
- infinite slew rate

> **Note**
>
> Impedance is similar to resistance but takes into account the effect of capacitors and inductors in an a.c. circuit.

Real op-amps

In practice, the open-loop gain is not infinite but it can be as high as 100 000.

The input impedance varies but is typically in excess of $10^6 \Omega$. A high input impedance reduces the effect of the internal resistance of the input supply.

The output impedance is less than 100Ω. This has to be low to reduce the fall in voltage as current is supplied from the output.

For an ideal amplifier, all frequencies should be amplified by the same factor in order to get a faithful reproduction of the input signal. Bandwidth is the range of frequencies which is amplified in this way.

In practice, an op-amp on open loop has a very small bandwidth.

Slew rate is a measure of how fast the output changes with respect to the input. Ideally the two would change simultaneously. However, this can lead to unwanted oscillations, so a terminal capacitance is included which limits the speed of the change at the output, the slew rate.

The op-amp as a comparator

An op-amp is a **differential amplifier** or **comparator**. This means that it amplifies the difference between the signals to the two inputs.

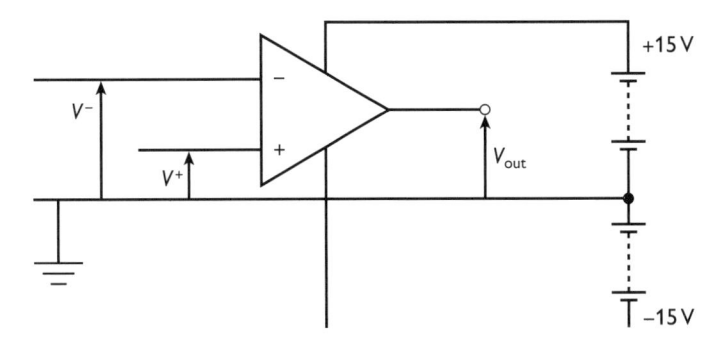

An op-amp being used as comparator

For an op-amp with input voltages V^+ and V^- and output voltage V_{out}, the open-loop gain G_0 is:

$G_0 = V_{out}/(V^+ - V^-)$

This means that if the strain gauge in the worked example on page 223 were connected to the non-inverting input, and the output from a potential divider with resistors of 50Ω and 150Ω were connected across the inverting amplifier of an op-amp of open-loop gain 10 000, then the difference between the two inputs would be:

$V^+ - V^- = 6.78 - 6.75 = 0.03$ V

and $V_{out} = 0.03 \times 10\,000 = 300$ V

In practice, this would not happen. The amplifier would saturate and the output would be equal to the voltage of the power supply, 15 V.

Feedback

A fraction of the output voltage can be fed back to either the inverting input (negative feedback) or to the non-inverting input (positive feedback). Negative feedback is useful in amplification because although it reduces the gain, it gives a much wider bandwidth, produces much less distortion in the output signal and increases the stability of the output signal.

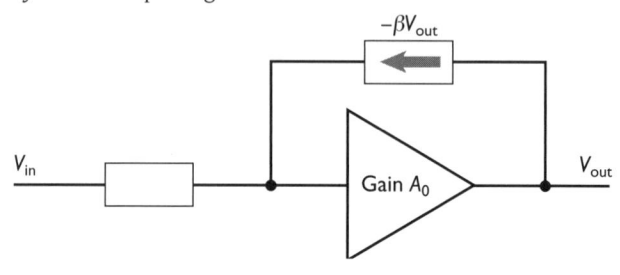

An op-amp with a fraction (β) of the output voltage fed back

In this diagram, a fraction (β) of the output voltage is fed back to the input. The minus sign is included because the output is 180° out of phase with the input, so the output is negative when the input is positive and vice versa.

So $V_{out} = A_0 \times$ (input to amplifier)

$V_{out} = A_0(V_{in} + \beta V_{out})$

$V_{out} - A_0\beta V_{out} = A_0 V_{in}$

Therefore the gain A of the amplifier in this configuration = V_{out}/V_{in}

$A = A_0/(1 + \beta A_0)$

Worked example

An operating amplifier has an open-loop gain of 10 000. Calculate the amplifier gain:

(a) when 50% of the output is fed back to the inverting input

(b) when 10% of the output is fed back to the inverting input

Answer

(a) $A = A_0(1 + \beta A_0) = 10\,000/(1 + 0.5 \times 10\,000) \approx 2$

(b) $A = A_0(1 + \beta A_0) = 10\,000/(1 + 0.1 \times 10\,000) \approx 10$

The inverting amplifier

There are two ways in which an op-amp can be used as an amplifier. In the first method the input voltage is connected to the inverting input, giving an output which is 180° out of phase with the input.

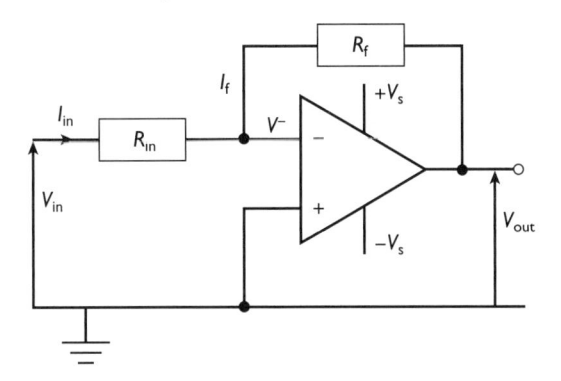

The inverting amplifier

Look at the diagram above. The non-inverting input is connected to earth. To avoid saturation, the voltage at the inverting voltage (V^-) must be nearly equal to the voltage at the non-inverting voltage. This is zero, hence V^- is almost at earth potential. This is known as the **virtual earth approximation**. It means that to a very close approximation:

$V_{in} = I_{in}R_{in}$

The input impedance of an op-amp is very high. Consequently, the input current to the amplifier is very small. Using Kirchhoff's first law, to a very close approximation:

$I_{in} = I_f$ where I_f is the current through the feedback resistor

Now consider the loop starting at earth, going through V_{in}, through R_{in} and R_f, through V_{out} and back to earth. Apply Kirchhoff's second law:

$V_{in} - I_{in}R_{in} - I_f R_f - V_{out} = 0$

but $V_{in} = I_{in}R_{in}$ and therefore:

$V_{out} = -I_f R_f$

gain $= V_{out}/V_{in} = -I_f R_f / I_{in}R_{in}$

but $I_{in} \approx I_f$ and therefore:

$A = -R_f/R_{in}$

> **Note**
>
> Making these approximations might appear to be an odd way of working but in practice it gives results that are well within any working tolerance. Electronics is a practical discipline, so this is all that is required.

The non-inverting amplifier

The second way to use an op-amp as an amplifier is with the input voltage connected to the non-inverting input. This gives an output that is in phase with the input.

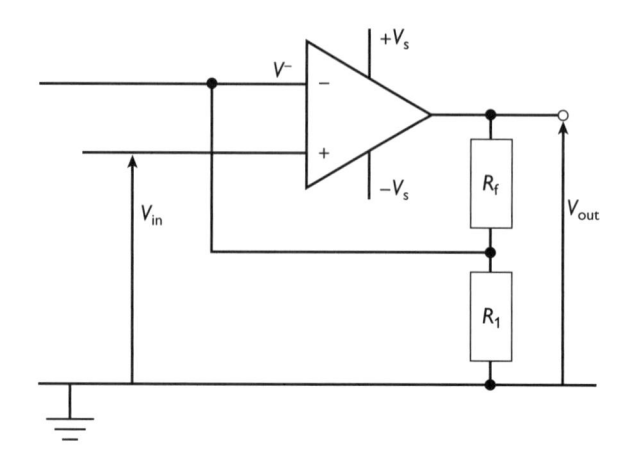

The non-inverting amplifier

In the circuit diagram above, you can see that the inverting amplifier is connected to the point between the two resistors R_f and R_1, thus providing the feedback. Assuming that the amplifier has not saturated, the overall gain = V_{out}/V_{in}:

$A = 1 + R_f/R_1$

Worked example 1

An inverting amplifier, with the circuit given in the diagram above, has a supply voltage of 15 V. The input resistance is 100 kΩ and the feedback resistance is 5 kΩ.

Calculate the output voltage when the input voltage is:
(a) 0.2 V

(b) 0.8 V

Answer
(a) $A = -R_{in}/R_f = 100/5 = -20$

  $V_{out} = AV_{in} = -20 \times 0.2 = -4$ V

(b) $V_{out} = AV_{in} = -20 \times 0.8 = -16$ V

The supply voltage is only 15 V, so the maximum output is 15 V. The amplifier will saturate at −15 V.

Worked example 2

Calculate the resistance required to make a non-inverting amplifier if the gain is to be 20 and the feedback resistance is 50 kΩ.

Answer

$A = 1 + R_f/R_1$

$20 = 1 + 50/R_1$

$R_1 = 50/19 = 2.6\,\text{k}\Omega$

The relay

Op-amps are often used in control circuits. For example, they can be used to switch a heater on when the temperature falls below a prescribed level. However, even if the output voltage is (theoretically) sufficient to drive the appliance, the output current from an op-amp is very small and is not large enough to do so. The op-amp can be used to drive a relay, which will switch the power circuit on.

The circuit diagram symbol for a relay is shown below. The rectangle represents the coil.

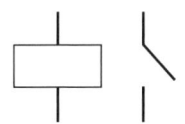

The relay is an electromagnetic switch. The op-amp drives a small current through a coil. This operates the switch, which connects/disconnects the power circuit.

Connecting a relay in a circuit

A relay cannot simply be connected to an op-amp. The relay is an electromagnetic switch and when it opens a large e.m.f. can be induced across the coil. This e.m.f. would destroy the op-amp.

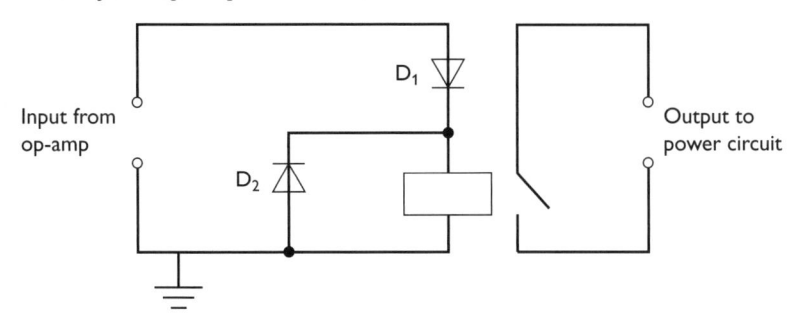

An op-amp output to a relay with protective diodes

In the diagram above, D_1 and D_2 are diodes. The output from the op-amp can be either positive or negative. D_1 allows current through the relay coil only when the output from the op-amp is positive. When the current is switched off, the induced e.m.f. is in the opposite direction (Lenz's law) and diode D_2 allows any current to pass harmlessly round the coil. Note that D_2 is connected so that when the input from the op-amp is positive, current will pass through the coil, not through the diode.

The light-emitting diode

Light-emitting diodes (LEDs), like any diode, allow current to pass through them in only one direction. When a current passes through an LED it emits light. LEDs have many uses and have major advantages over hot filament lights. Principally they are much more efficient at converting electrical energy to light and consequently take much less current.

LEDs are used as indicator lights in many op-amp circuits, as shown below.

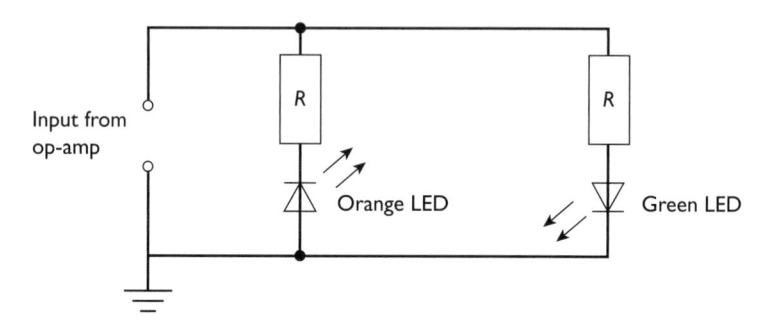

In this type of circuit, when the input from the op-amp is positive the green LED lights; when the input is negative the orange LED lights. This type of circuit could be used to indicate the state of charge of a charging battery. The potential difference across a resistor in the charging circuit and the potential difference across the battery are applied to the two inputs of an op-amp acting as a comparator. When the battery is uncharged there is a relatively large current through the resistor, the potential difference is larger than that across the battery and the output from the op-amp is negative (orange LED lights). As the battery charges, the current falls. Consequently, the potential difference across the resistor falls and that across the battery increases. When the battery is fully charged the potential difference across it is greater than that across the resistor. The op-amp output becomes positive and the green LED lights.

Calibration of meters

An op-amp can be used to monitor a changing physical quantity such as temperature. The output from the op-amp is unlikely to be linear and, if this is being monitored by a voltmeter connected across the output, it will not give a direct reading of the temperature. A **calibration curve** is required. The reading on the voltmeter is recorded at different known temperatures and a graph (a calibration curve) is plotted. The temperature can now be obtained by referring to the reading on the voltmeter and then using the graph.

Remote sensing

Advantages of remote sensing

This section looks at remote sensing in medicine. Until the discovery of X-rays at the beginning of the twentieth century, the only ways to diagnose a medical condition were from external symptoms or by cutting the patient open. The discovery of X-rays allowed doctors to see internal organs without surgery. Clearly, this is much less traumatic for the patient and reduces significantly the chances of infection from open wounds. Today, X-rays are only one method in a whole range of non-invasive diagnostic techniques used in medicine.

Production of X-rays

X-rays are formed when electrons are accelerated to very high energies (in excess of 50 KeV) and targeted onto a heavy metal. Most of the electron energy is converted to heat but a small proportion is converted to X-rays.

In the diagram below, the cathode is shown. This will be heated indirectly. When a metal is heated to a high temperature, electrons are emitted from the surface of the metal. This is known as **thermionic emission**. The anode is at a much higher potential than the cathode, consequently the electrons emitted by the cathode are accelerated towards it at very high speeds.

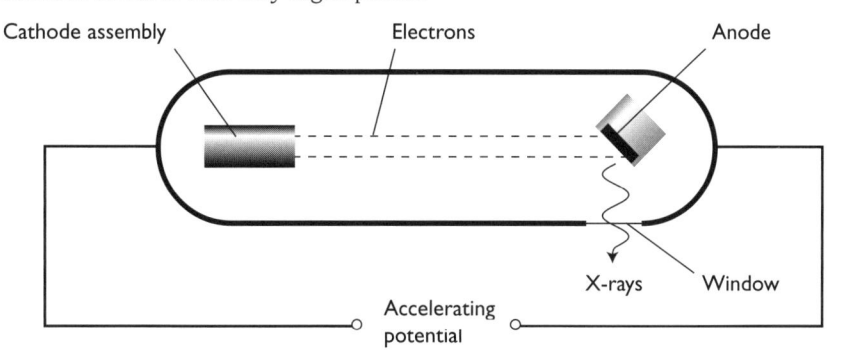

Principle of X-ray production

Table 20 shows some of the features of a modern X-ray tube.

Table 20

Modification	Reason
Rotating anode	To avoid overheating of the anode
Coolant flowing round anode	To avoid overheating of the anode
Thick lead walls	To reduce radiation outside the tube
Metal tubes beyond the window	Collimate and control the width of the beam
Cathode heating control	The cathode in a modern tube is heated indirectly; the current in the heater determines the temperature of the cathode

The intensity of the X-ray beam is controlled by the number of electrons hitting the anode per unit time. This is the **tube current**. The greater the rate of arrival of electrons the greater is the intensity. The tube current is controlled by the rate of emission of electrons from the cathode. This in turn is controlled by the temperature of the cathode. The higher the temperature, the greater is the rate of emission of electrons.

Reducing the dose

A wide range of X-ray frequencies is emitted from a simple X-ray tube. The very soft (long) wavelength rays do not penetrate through the body of the patient, yet would add to the total dose received. An aluminium filter is used to absorb these X-rays before they reach the patient.

X-ray spectrum

A typical X-ray spectrum consists of a line emission spectrum superimposed on a continuous spectrum, as shown in the diagram below.

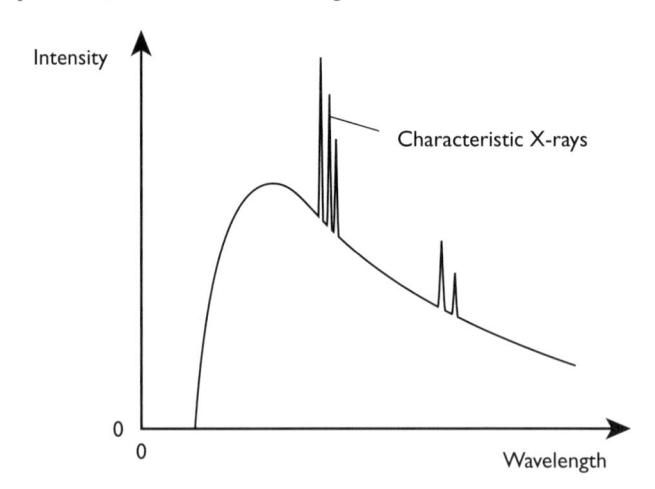

A typical X-ray spectrum

The continuous part of the spectrum is known as Bremsstrahlung or braking radiation. It is caused by the electrons interacting with atoms of the target and being brought to rest. Whenever charged particles are accelerated electromagnetic radiation is emitted. The large decelerations involved in bringing the electrons to rest produce high-energy photons that are in the X-ray region. Sometimes the electrons lose all their energy at once, and at other times give it up through a series of interactions — hence, the continuous nature of the spectrum. The maximum photon energy cannot be greater than the maximum energy of the incident electrons. This means that there is a sharp cut off at the maximum frequency/minimum wavelength of the X-rays, (remember the photon energy is directly proportional to the photon frequency). This cut-off point is determined by the accelerating potential.

The **hardness** of the X-rays is a measure of their penetrating effect. The higher the frequency of X-rays, the greater is their penetration. Thus the hardness of X-rays is also determined by the accelerating voltage.

The characteristic X-rays and their energies are dependent on the target metal; they are of little importance in medical physics. These are line emission spectra. The incident electrons excite the inner electrons of the anode material up to higher energy levels. When an excited electron drops back to the ground state, a photon of a specific energy, and hence frequency, is emitted.

If you refer back to the diagram on page 232 and remember that the photon energy is equal to Planck's constant times the frequency, then you can see that the frequency–intensity graph will have an identical shape.

> **Worked example**
> Calculate the minimum wavelength of X-rays that can be produced when electrons are accelerated through 120 kV.
>
> **Answer**
> energy of electrons $= eV = 1.6 \times 10^{-19} \times 120 \times 10^3 = 1.92 \times 10^{-14}$ J
>
> $E = hf = hc/\lambda$
>
> Therefore:
> $\lambda = hc/E = 6.63 \times 10^{-34} \times 3.0 \times 10^8/(1.92 \times 10^{-14}) = 1.04 \times 10^{-11}$ m

Uses of X-rays in medical imaging

The major use of X-rays is in diagnostics, particularly for broken bones and ulcerated tissues in the duodenum and other parts of the gut. Bone tissue is dense and is a good absorber of X-rays; flesh and muscle are much poorer absorbers. Therefore, if a beam of X-rays is incident on an area of damaged bone a shadow image is formed. The bones appear light, because very few X-rays reach the film. The background, where many X-rays reach the film, will be much darker. It is worth noting that this is a 'negative' image.

When considering the diagnosis of ulcers there is little difference in the absorption by healthy tissue and ulcerated tissue. In order to improve the **contrast**, the patient is given a drink containing a salt that is opaque to X-rays. The ulcerated tissue absorbs more of the salt than healthy tissue and hence absorbs much more of the X-radiation. This material, often a barium salt, is called a **contrast medium**.

Clarity of images
The clarity or clearness of an image depends on the contrast between the dark areas and the light areas and on the sharpness of the image. Contrast is the difference between the dark and lighter areas of the image. Table 21 gives the main methods by which contrast can be improved.

VII Gathering and communicating information

Table 21

Method	Detail
Contrast medium	Used to distinguish between soft tissues
Longer exposure time	This improves contrast but has the disadvantage of increasing the patient's exposure to X-radiation
Choice of X-ray hardness	Much harder (more penetrative) X-rays are used for investigating bone injuries than for soft-tissue diagnosis, e.g. stomach ulcers
Image intensifier	At its simplest, an image intensifier can be a fluorescent sheet placed at the back of the photographic film. X-rays that pass through the film hit this, causing it to fluoresce. The detector picks up this, making the dark parts darker. More sophisticated image intensifiers consist of cells in which the incident X-rays liberate electrons in a photocathode. These are fed to a digital detector

The sharpness of an image is determined by the width of the incident beam and its collimation (how parallel it is). The narrower the beam, the sharper is the image.

The cross-section of the beam depends on:
- the size of the anode — the larger the anode, the greater the width of the beam
- the diameter of the window — the larger the window, the greater the width of the beam

Likewise, the better the collimation the sharper is the image. This can be improved by passing the beam through narrow slits as shown in the diagram below.

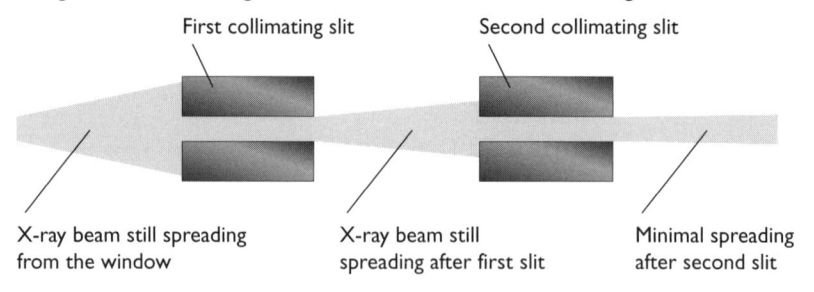

First collimating slit   Second collimating slit

X-ray beam still spreading from the window   X-ray beam still spreading after first slit   Minimal spreading after second slit

Scattering

Some X-rays are scattered from various organs in the body. If these reach the plate on which the image falls they lead to a loss of sharpness. To avoid this happening, an anti-scatter grid is used. This is a series of parallel aluminium and lead plates, as shown in the diagram below.

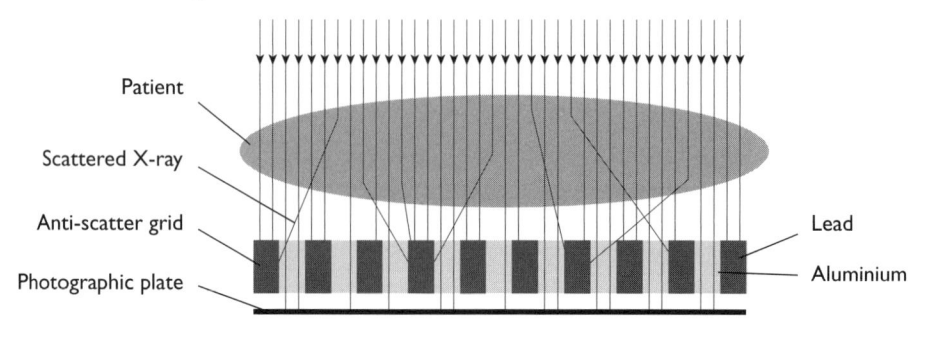

Patient

Scattered X-ray

Anti-scatter grid

Photographic plate

Lead

Aluminium

The aluminium allows the X-rays through, whereas the lead absorbs any scattered X-rays.

Attenuation of X-rays

The formal term for the decrease in intensity of a signal as it passes through a material is **attenuation**. Attenuation of X-rays depends on the material it is passing through — for dense materials such as bone it is high; for less dense material such as flesh it is much lower. However, for each, the attenuation (provided the beam is parallel) is exponential in nature giving the equation:

$$I = I_0 e^{-\mu x}$$

where I is the intensity, I_0 is the initial intensity, μ is the linear attenuation (or absorption) coefficient and x is the thickness of material the signal has passed through.

The mathematics of, and dealing with, this equation are identical to that for the radioactive decay equation. The equivalent of half-life is the **half-value thickness** (h.v.t.) of the material. This is the thickness of material that reduces the intensity of the incident signal to half of its original intensity.

The linear attenuation coefficient depends not only on the material but also on the hardness of the X-rays that are used.

> **Worked example**
>
> Bone has a linear attenuation coefficient of $0.35\,\text{cm}^{-1}$ for X-rays of a particular frequency.
>
> Calculate the half-thickness of bone for this type of X-ray.
>
> **Answer**
> $x_{1/2} = \ln 2/\mu = 0.693/0.35 = 2.0\,\text{cm}$

Computerised tomography scanning

The traditional use of X-rays has the major disadvantage of producing only a shadow image. This makes it difficult to identify the true depth of organs. Computerised tomography (CT) takes the technology of X-rays a step further. The main principles of CT scanning are:
- The patient lies in the centre of a ring of detectors. An X-ray source moves around the patient taking many images at different angles.
- The images are put together, using a powerful computer, to form an image of a slice through the patient.
- The patient is moved slightly forward so an image of another slice is made. This is repeated for many slices.
- The computer puts the slices together to form a three-dimensional image that is rotated so that medical practitioners can view the image from different angles.

VII Gathering and communicating information

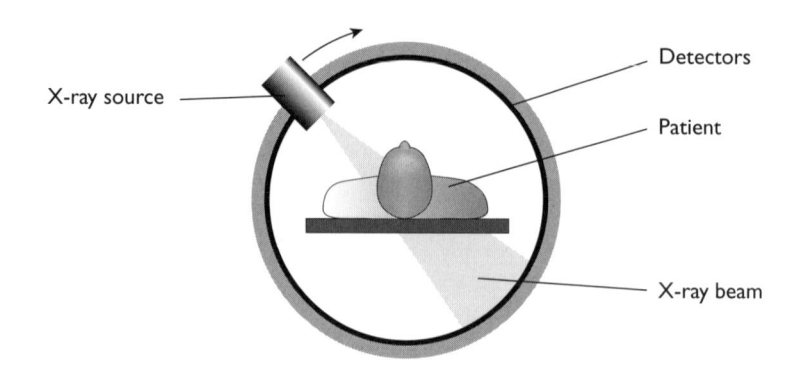

Build up of the image

The body part under investigation is split into tiny cubes called **voxels**. The build-up of the picture depends on voxels having different linear absorption coefficients, thus allowing different signal strengths to reach the detector. As the camera is moved round the circle, the signal strengths from different directions are measured and an image is built up. The intensity of the beam is reduced as it passes through each voxel. This reduction gives the **pixel intensity** (w, x, y and z).

Step 1
X-rays are incident on the voxel cube. The detector adds pixel intensities receiving a total of $w + x$ on the upper layer and $y + z$ on the lower layer, interpreting this as an array:

$w + x$	$w + x$
$y + z$	$y + z$

Step 2
The X-ray tube swings round. When the X-ray source is at 45°, the detector interprets this as an array:

w	$x + y$
$x + y$	z

This is then added to the array in Step 1 to give:

$2w + x$	$w + 2x + y$
$x + 2y + z$	$y + 2z$

Step 3
Now consider the position when the X-ray tube has swung round a further 45°. The detector interprets this as an array:

$w + y$	$x + z$
$w + y$	$x + z$

This is then added to the array in Step 2 to give:

$3w + x + y$	$w + 3x + y + z$
$w + x + 3y + z$	$x + y + 3z$

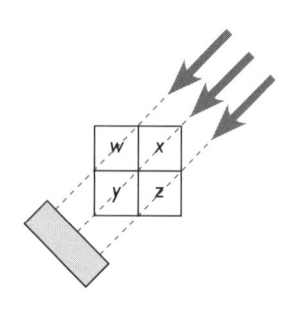

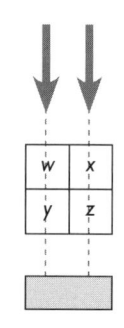

236

Step 4
When the source swings round a further 45° the detector interprets this as an array:

w + z	x
y	w + z

Adding this to the final array in Step 3 gives:

4w + x + y + z	w + 4x + y + z
w + x + 4y + z	w + x + y + 4z

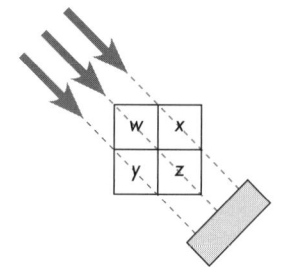

Step 5
The background intensity is subtracted from the totals — this is the sum of the intensities at any one position, in this case: w + x + y + z.

minus:

w + x + y + z	w + x + y + z
w + x + y + z	w + x + y + z

gives:

3w	3x
3y	3z

Step 6
This is then divided by 3 because the process is repeated at three more angles. This gives the original figures:

w	x
y	z

Try working through this procedure using the initial values 5, 6, 1 and 8.

Building up a CT image

This calculation is greatly simplified. In reality, there are far more voxels and the scanner takes readings at hundreds of angles to produce a picture of just one slice. In addition to this, X-rays of different hardness are used in order to identify different types of tissue. You can begin to see the necessity for the use of a powerful computer to do the calculation, and you may not even have started to think about combining the many slices to form a three-dimensional image!

Ultrasound

Ultrasound waves are sound waves that have frequencies above the threshold of human hearing, which is 20 kHz. Pulses of ultrasound are directed to the organs under investigation. Different percentages of incident ultrasound reflect off the boundaries of different types of tissue, enabling a 'sound' picture to be built up.

As with any waves, the resolution that can be obtained is limited due to diffraction effects. Very high frequency waves are used in medical ultrasound scanning in order to resolve small details.

Worked example

The approximate speed of sound in human tissue is about 1500 m s⁻¹. Estimate the minimum frequency ultrasound that would enable a doctor to resolve detail to the nearest 0.1 mm.

Answer

To resolve detail to the nearest 0.1 mm, the wavelength of the ultrasound must be of the order of 0.1 mm.

$f = c/\lambda = 1500/(0.1 \times 10^{-3}) = 1.5 \times 10^7 \, Hz = 15 \, MHz$

Production of ultrasound

The transducer that produces and receives the ultrasound waves relies on the piezo-electric effect (see page 223). A short pulse of high frequency alternating voltage input causes the crystal to vibrate at the same frequency as the input voltage, producing an ultrasonic wave pulse. The ultrasonic wave pulse is reflected back from different tissues and is received by the same crystal. This causes it to vibrate and induce e.m.f.s that are sent to a computer. Thus the same crystal transducer is both the ultrasound generator and detector. This is illustrated in the diagram below.

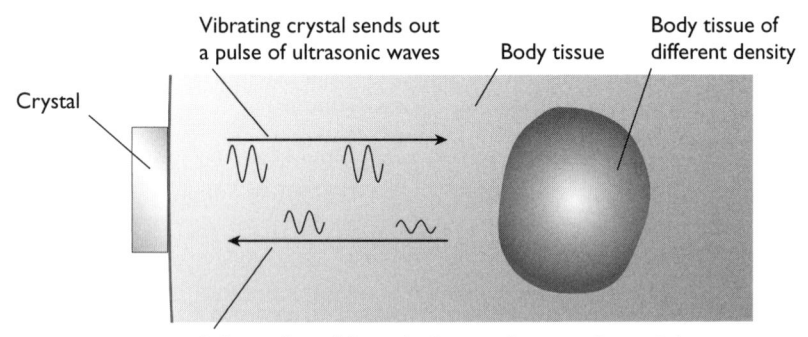

Crystal — Vibrating crystal sends out a pulse of ultrasonic waves — Body tissue — Body tissue of different density

Pulses reflected from the front and rear surfaces of the denser material, which in turn cause the crystal to vibrate

As with many systems, resonance is required for maximum efficiency — the frequency of the ultrasound should be equal to the natural frequency of vibration of the crystal. For this to occur, the thickness of the crystal must be equal to one-half of the ultrasound wavelength.

Note

This should remind you of the work on the formation of stationary waves.

Worked example

The speed of ultrasound in a piezoelectric crystal is 3600 m s⁻¹. Calculate the thickness of the crystal that would be used to generate waves of frequency 4.0 MHz.

Answer

$\lambda = c/f = 3600/(4.0 \times 10^6) = 9.0 \times 10^{-4}$ m

thickness of the crystal $= \frac{1}{2}\lambda = 4.5 \times 10^{-4}$ m $= 0.45$ mm

Reflection of ultrasound

When ultrasound moves from a material of one density to another, some of the ultrasound is refracted and some is reflected. The fraction reflected depends on the **acoustic impedance**, Z, of the two materials:

$Z = \rho c$

where ρ is the density of the material and c is the speed of the ultrasound in the material.

Worked example

The density of blood is 1060 kg m^{-3} and the speed of ultrasound in it is 1570 m s^{-1}. Calculate the acoustic impedance of blood.

Answer

$Z = \rho c = 1060 \times 1570 = 1.66 \times 10^6$ kg m^{-2} s^{-1}

The fraction of the intensity of the ultrasound reflected at the boundary of two materials is calculated from the formula:

$$\frac{I_r}{I_0} = \frac{(Z_1 - Z_2)^2}{(Z_1 + Z_2)^2}$$

where I_r is the intensity of the reflected beam, I_0 is the intensity of the incident beam and Z_1 and Z_2 are the acoustic impedances of the two materials.

The ratio I_r/I_0 indicates the fraction of the intensity of the incident beam reflected and is known as the **intensity reflection coefficient**.

Note

This equation is really only accurate for angles of incidence of 0° but it gives a good approximation for small angles.

Worked example

The density of bone is 1600 kg m^{-3} and the density of soft tissue is 1060 kg m^{-3}. The speed of sound in the two materials is 4000 m s^{-1} and 1540 m s^{-1} respectively.

Calculate the intensity of the reflected beam compared with the incident beam.

Answer

$Z_{bone} = 1600 \times 4000 = 6.40 \times 10^6 \, kg\,m^{-2}\,s^{-1}$

$Z_{soft\,tissue} = 1060 \times 1540 = 1.63 \times 10^6 \, kg\,m^{-2}\,s^{-1}$

$$\frac{I_r}{I_0} = \frac{(Z_1 - Z_2)^2}{(Z_1 + Z_2)^2} = \frac{(6.40 - 1.63)^2}{(6.40 + 1.63)^2} = \frac{4.77^2}{8.03^2} = 0.35$$

Note

The arithmetic is simplified by ignoring the factor 10^6. It is common to all terms in the equation and therefore cancels.

Coupling medium

The speed of sound in air is approximately $300\,m\,s^{-1}$ and the density of air is about $1.3\,kg\,m^{-3}$. This gives an acoustic impedance of approximately $400\,kg\,m^{-2}\,s^{-1}$. Comparing this with skin, it means that 99.9% of the incident wave would be reflected at the air–skin boundary. To avoid this, a gel with a similar acoustic impedance to that of skin is smeared on the skin and the ultrasound generator/receiver is run across this. This gel is known as a **coupling agent**.

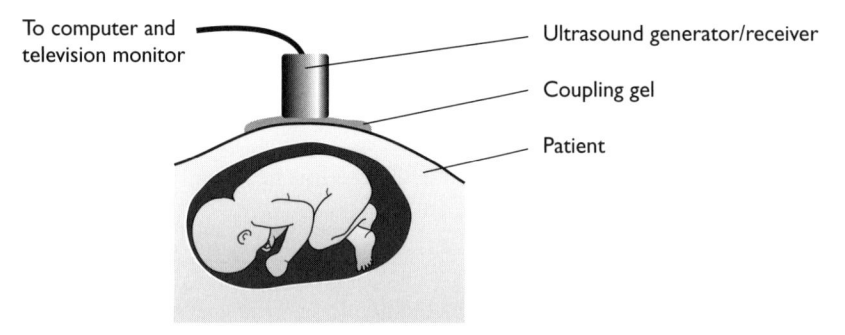

A-scan

A pulse of ultrasound is passed into the body and the reflections from the different boundaries between different tissues are received back at the transducer. The signal is amplified and then displayed as a voltage–time graph on an oscilloscope screen.

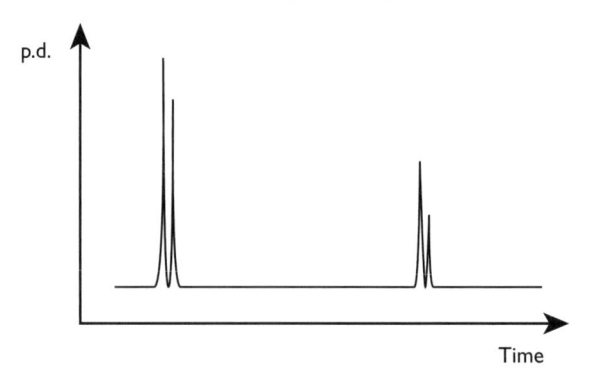

Oscilloscope display of ultrasound pulses

The graph above might show the reflections from the front and back of a baby's skull. There are two reflections at each surface — one from the outer part of the skull bone and one from the inner part. Such a scan would give evidence of both the thickness and the diameter of the skull.

Worked example
Ultrasound travels at a speed of $1500\,\text{m s}^{-1}$ through brain tissue. In the graph above, the time base of the oscilloscope is set at $50\,\mu\text{s div}^{-1}$.

Calculate the diameter of the baby's skull.

Answer
separation of the two peaks = 4 divisions = $4 \times 50 = 200\,\mu\text{s}$

distance the ultrasound pulse travels = $vt = 1500 \times 200 \times 10^{-6} = 0.30\,\text{m}$

diameter of the baby's skull = $\tfrac{1}{2} \times 0.30 = 0.15\,\text{m}$

Note
The signal has to travel across the gap between the two sides of the skull, is then reflected and travels back the same distance. Hence, the distance calculated from the graph is twice the diameter of the skull.

It is difficult to decide exactly where to take the readings from on each peak. in the example given, the readings are taken from the inner surface of the skull.

Problems include:
- The pulses received from reflections deeper in the body are weaker than those reflected from boundaries less deep in the body.
- The ultrasound waves are scattered and absorbed to some extent as they travel through the body.
- The reflected pulse is reflected again at the different boundaries as it travels back to the transducer.

In order to overcome these problems, the later pulses are amplified more than the ealier pulses.

B-scan
A B-scan uses a similar technique but the generator/receiver is moved around so that readings are taken from different angles. The position of the transducer is mapped in a similar way to a computer mouse recording its position on a mouse pad. This, and the information from all the different angles (both time lags and reflection intensities), are sent to a computer. Each pulse is then represented on the screen as a bright dot, which builds up a two-dimensional image. In practice, B-scan probes consist of an array of many transducers each of which is at a slightly different angle. This reduces the time it takes to build up an image and, therefore, reduces blurring.

The advantage of using ultrasound rather than X-rays is that the patient and staff are not receiving potentially harmful, ionising radiation.

Attenuation of ultrasound

In a similar way to X-rays, ultrasound is absorbed as it passes through tissue. The amount of absorption, as with X-rays, depends on both the type of tissue and on the frequency of the wave. The mathematics of X-ray and ultrasound attenuation are identical although, of course, the attenuation (or absorption) coefficients are different.

$$I = I_0 e^{-kx}$$

where I_0 is the initial intensity of the beam, I is the intensity of the beam after passing through a material of thickness x and k is the acoustic absorption coefficient.

Magnetic resonance imaging

Theory of nuclear magnetic resonance

The nuclei of some atoms have a property called spin — you can imagine them as tiny spinning tops. Because the nuclei are charged, this spin makes them behave like tiny magnets. When in a magnetic field they do not spin exactly parallel to the field but 'wobble', or precess, about the direction of the field, as shown in the diagram below.

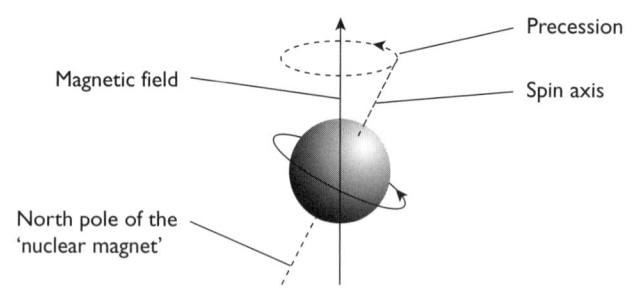

The angular frequency of the precession is known as the **Larmor frequency**. It depends on the particular type of nucleus and is proportional to the strength of the flux density of the applied magnetic field.

In living tissue there are large amounts of water, and hence hydrogen atoms. It is the hydrogen nuclei (protons) that are used in MRI scans.

In general, nuclei spin so that they are in the lowest energy state, with the tiny magnets as in the diagram. A few, however, move into a higher energy state with the magnet reversed. Nuclear magnetic resonance relies on encouraging more nuclei to flip into the higher energy state. This is done by superimposing on top of the strong permanent field, a field that oscillates at the same frequency as the frequency of precession, which is in the radio frequency (RF) region. This is **nuclear magnetic resonance**.

Once the RF field is switched off, the nuclei drop back into the lower energy state. This is called **relaxation**. When the atom drops back into the lower energy state in this way, a photon of energy is released. This is detected by the same coils that previously supplied the RF frequency field. The time the atoms take to relax (the **relaxation time**) is dependent on the type of tissue in which the atom occurs. Watery tissues have long relaxation times; fatty tissues have much shorter relaxation times. From the different relaxation times, a picture of the patient's internal organs can be built up.

MRI scanning procedure

In an MRI scan, the patient lies on a bed surrounded by large coils that supply the uniform field. As well as this uniform field, gradient coils produce a non-uniform field that varies across the length, width and depth of the patient. This means that the resonant frequency is slightly different for each small part of the body. This can then be identified by the computer.

The RF field is applied in pulses. The patient moves slowly through the coils with a set of information taken at each position. This information is sent to a computer, which builds up an image.

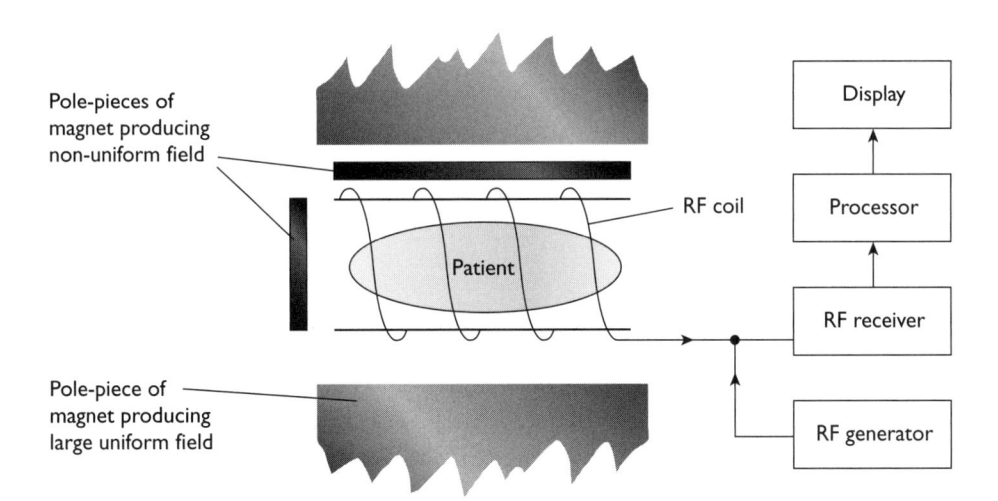

The principles of a magnetic resonance scanner

Advantages and disadvantages of MRI

Advantages of MRI include:
- no ionising radiation
- good contrast between different soft tissues
- three-dimensional images or 'slices' can be produced
- no side-effects for patients

Disadvantages of MRI include:
- It is not suitable for patients with pacemakers or with metallic replacement joints or surgical pins in the body.
- The patient must lie very still during the procedure.

Communicating information

Modulation

Electromagnetic radiation, mostly in the radio wave and microwave regions of the spectrum, is used to transmit information across long distances. The information is transmitted by a wave of specific frequency to which a receiver is tuned. This wave is known as the **carrier wave**. When you change stations on a radio receiver you are tuning in to a different carrier frequency. The information is carried by modulating the carrier wave. There are two forms of **modulation**; **amplitude modulation** and **frequency modulation**.

Amplitude modulation

The carrier wave has a constant frequency; the amplitude varies in synchrony with the displacement of the information signal.

The variation of amplitude is detected by the receiver and converted back into a sound wave.

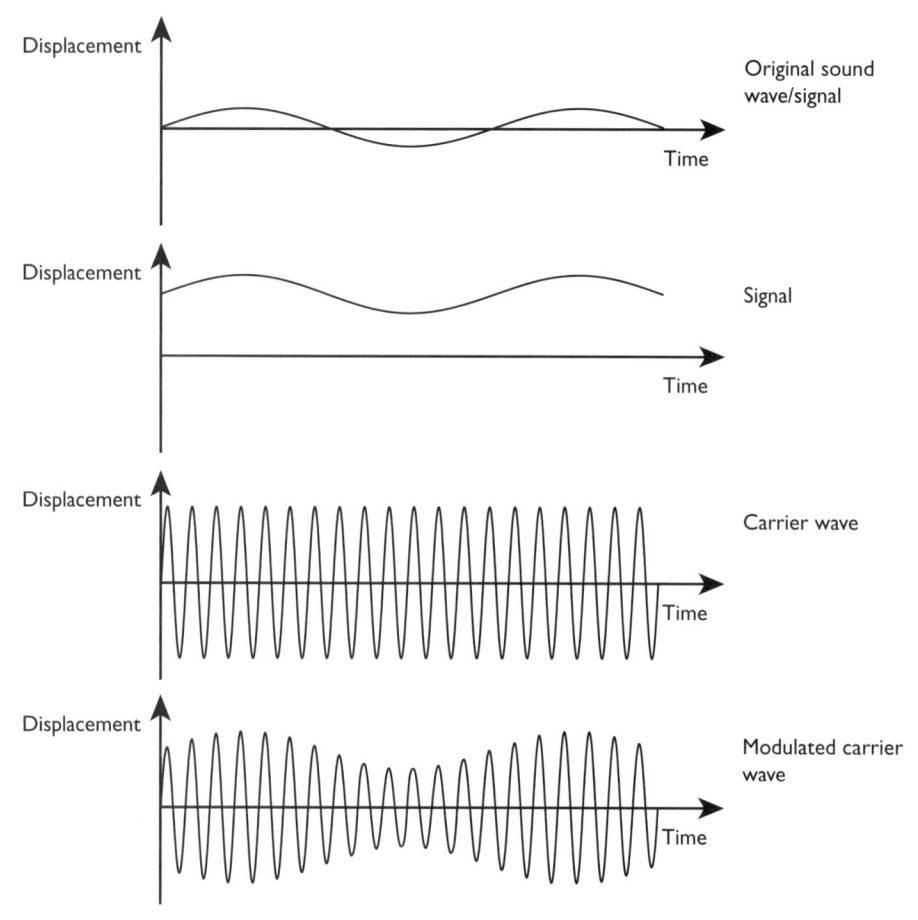

Amplitude modulation

Side bands and bandwidth

A carrier wave that is amplitude modulated by a signal of a single frequency consists of the carrier wave and two **sideband** frequencies.

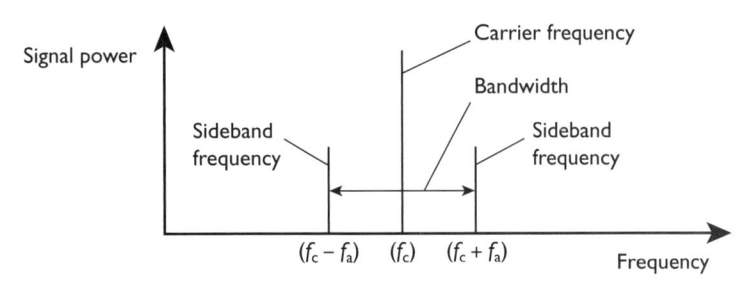

Sidebands with a wave modulated with a single frequency signal

If the frequency of the carrier wave is f_c and the frequency of the signal is f_s, then the frequencies of the two side bands are $f_c - f_s$ and $f_c + f_s$. In practice, when listening to a radio, there is a range of sound frequencies and, therefore, signals. The bandwidth is the range of these frequencies.

Bandwidth = $2f_s$

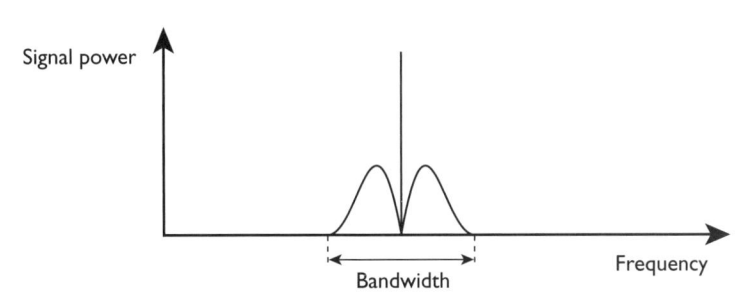

Sidebands with a wave modulated by a signal with a range of frequencies

Worked example

A radio station transmits at a frequency of 200 kHz. The maximum bandwidth is 9 kHz.

Calculate the lowest and highest frequencies that are transmitted. Comment on the effect on the reproduction of sounds.

Answer

highest frequency = $f_c + f_s$ = 200 + (9/2) = 204.5 kHz

lowest frequency = $f_c - f_s$ = 200 − (9/2) = 195.5 kHz

The maximum frequencies that the human ear can detect are between 15 kHz and 20 kHz. With a bandwidth of only 9 kHz the highest notes are lost and the quality of sound reproduction is reduced.

Frequency modulation

With frequency modulation, the amplitude remains constant but the frequency of the carrier wave is varied in synchrony with the information signal.

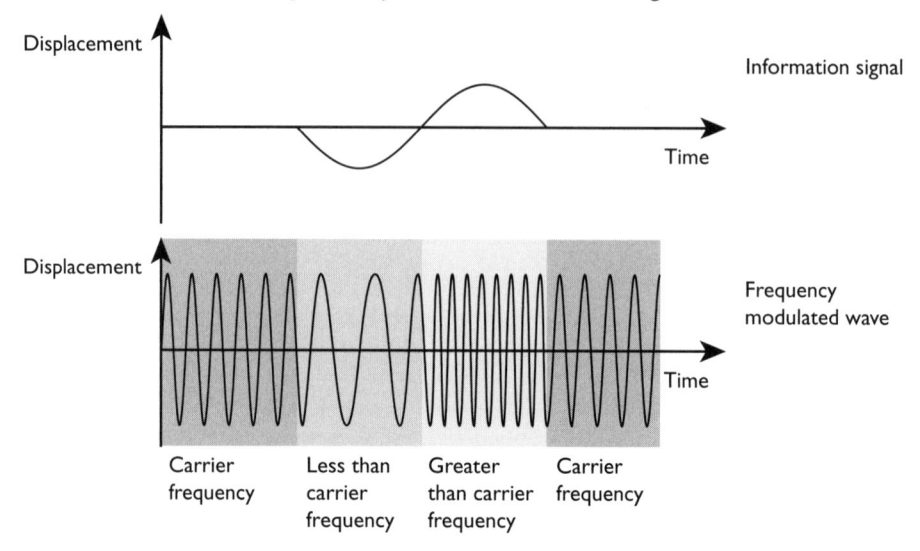

Frequency modulation

FM broadcasts have a wider bandwidth than AM waves — 15 kHz compared with 9 kHz. Although some humans can detect sounds of up to 20 kHz, these sounds tend to be very faint, so this bandwidth is capable of high-quality reproduction of music.

Advantages of AM and FM

AM and FM are compared in Table 22.

Table 22

Amplitude modulation	Frequency modulation
Narrower bandwidth, more stations available in any frequency range	Wider bandwidth means better reproduction of sound
Long wavelength means the waves can be diffracted round physical barriers; particularly significant in mountainous areas; fewer transmitters are required.	FM cannot be diffracted round physical barriers; less prone to interference from other sources, e.g. lightning, unsuppressed internal combustion engines
Range of AM transmissions is much bigger than FM; hence this also means fewer transmitters.	Range of FM transmissions is much smaller than AM
Cheaper and less advanced technology required for transmission and receiving	More energy efficient; in AM waves one-third of the total power is carried by the sidebands
Can reflect off the ionosphere – enabled first transatlantic broadcast to be made – not so relevant today with communications satellites	Constant amplitude in FM means constant power, unlike AM

Analogue and digital signals

The signals described so far have been analogue signals. A precise electrical image of the original sound waves is formed. This is similar to the way that when a microphone is connected to a cathode-ray oscilloscope a graphical picture of the original sound wave is seen. An analogue quantity can have any value.

Digital quantities are quite different. They can only have a series of set values. A digital signal is a series of voltage pulses — on/ off or high/low or 1/0.

An analogue signal is sampled at regular time intervals to build up a picture of the wave prior to converting to a digital signal. This is shown in the diagram below.

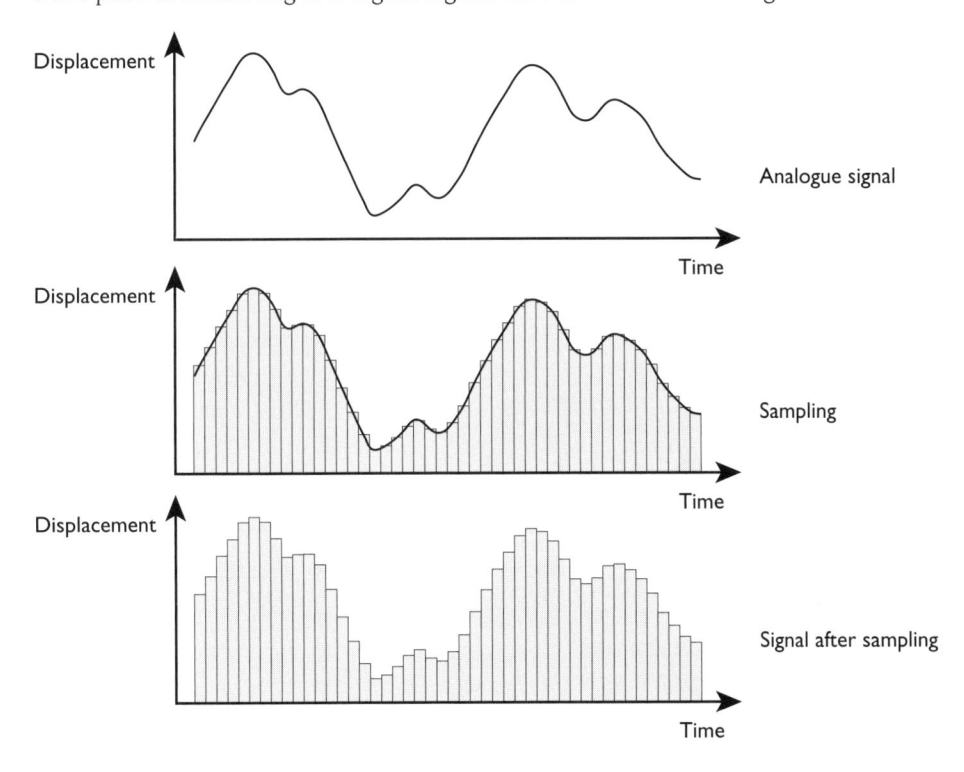

A digital system counts in **binary**. The precision of the measurement of the strength of a signal depends on the number of **bits**. A bit is a digit in the binary system. A four-bit system splits the signal into one of 16 levels. The quality of reproduction depends on:
- the number of bits employed in the system — the more bits, the smaller the steps in the rebuilt wave
- the frequency of sampling — the higher the frequency, the more often the wave is sampled and the better the quality.

In the diagram above, you can see that, although the overall shape of the signal can be determined, much definition is lost. More frequent sampling would give a more faithful reproduction. In order to get a faithful reproduction of a signal, the frequency

VII Gathering and communicating information

of sampling must be at least twice the maximum frequency of the signal. The human ear can detect notes of frequencies up to about 15–20 kHz. If a piece of music is to be reproduced accurately the sampling rate must therefore be at least 40 kHz. This is expensive to do and requires sophisticated technology. Consequently, telephone systems use a much lower sampling rate (about 8 kHz) because frequencies of about 3.5 kHz only are required for voice recognition.

> **Note**
>
> **Binary** is a system of counting to base 2, whereas we normally use base 10. Binary has the advantage that there are only two digits, 1 and 0. These are called bits.
>
> A four-bit system can count up to 15 whereas an eight-bit system can count up to 255. An eight-bit electronic system could split a signal into 256 levels.
>
> | Decimal | Binary |
> |---|---|
> | 0 | 0000 |
> | 1 | 0001 |
> | 2 | 0010 |
> | 3 | 0011 |
> | 4 | 0100 |
> | 5 | 0101 |
> | 6 | 0110 |
> | 7 | 0111 |
> | 8 | 1000 |
> | 9 | 1001 |
> | 10 | 1010 |
> | 11 | 1011 |
> | 12 | 1100 |
> | 13 | 1101 |
> | 14 | 1110 |
> | 15 | 1111 |

When information is transmitted, it is sent as a series of pulses. Each pulse contains the binary code for that piece of information. The device that converts the analogue signal into the binary (digital) number is called an **analogue-to-digital converter (ADC)**.

> **Worked example**
>
> A four-bit analogue-to-digital converter converts a signal into binary code. The ADC works at a maximum potential difference of 7.50 V.
>
> (a) Calculate the size of each 'step' in the output signal. Assume that there is no amplification or attenuation.
>
> (b) Deduce the binary code packets that would correspond to the following series of sampling voltages: 0.2 V, 1.5 V, 4.5 V, 7.2 V

International AS and A Level Physics Revision Guide

> *Answer*
> **(a)** voltage steps = 7.5/15 = 0.5 V
>
> **(b)** 0.2 V is below the first step, so is recorded as 0000
>
> 1.50 V is the third step (1.50/0.5 = 3) and is recorded as 0011
>
> 4.50 V is the ninth step (4.50/0.5 = 9) and is recorded as 1001
>
> 7.2 will read as 7.0 V, which is the fourteenth step and is recorded as 1110

At the receiver, the digital signal is converted back into an analogue signal. The device that does this is called a **digital-to-analogue converter (DAC)**.

Advantages of using digital signals
- Digital signals acquire much less noise than analogue signals through regeneration amplifiers.
- Noise, as well as the required signals, is amplified in analogue systems.
- Modern digital circuits are cheaper to manufacture than analogue systems.
- The bandwidth is smaller so more information can be transmitted per unit time.
- Extra information (extra bits of data) can be added to digital signals by the transmission system to check for any errors caused by the transmission.
- They are easier to encrypt for secure transmission of data.

Look at the first two points. It is inevitable that signals, not only radio signals but signals sent down wires or fibre-optic cables, will pick up noise as they travel long distances. They will also attenuate (get weaker). They will tend to smear out as the different frequencies in the signal travel at slightly different speeds along a cable or with slightly different paths.

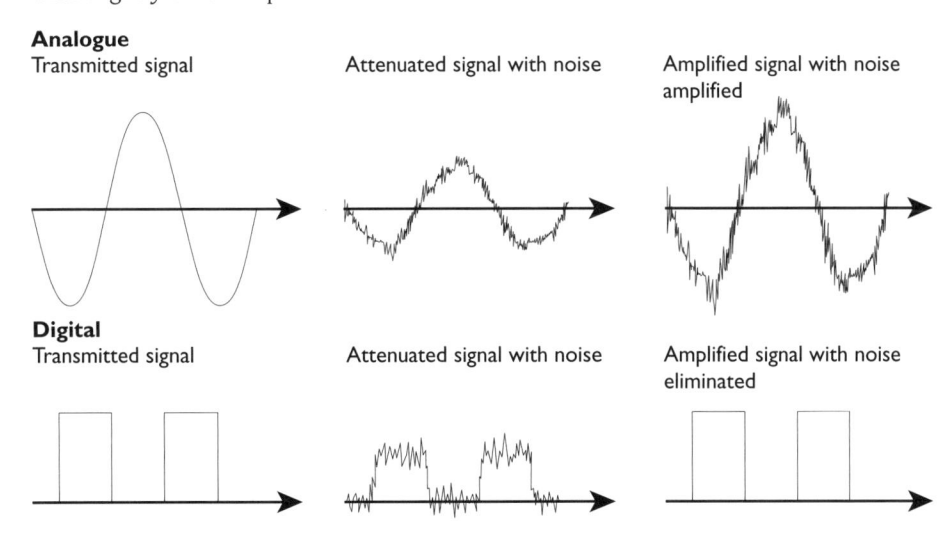

Comparison of amplification of analogue and digital signals

To counter the effects of attenuation, regeneration amplifiers amplify the signals at regular intervals. The noise on an analogue system is amplified along with the required signal. The noise with the digital amplification is eliminated because it is below the minimum step change that the amplifier recognises.

Transmission channels

Wire pairs
The first electrical method of transmission of information was the telegraph. It consisted of a transmitter connected by copper wires to a receiver in the form of a buzzer. Morse code signals were used to transmit the information. Some telephone systems still use copper wires but these have significant disadvantages and are being replaced gradually.

The disadvantages include:
- **cross-linking** — the signal intended for one subscriber is picked up by another, unintended, subscriber. It is caused by the transmitted signal on one circuit inducing a copy of the signal into an adjacent circuit.
- **poor security** — it is easy to tap into a telephone conversation.
- **high attenuation** — the electrical resistance of the wires means that the signal weakens relatively rapidly. The wires themselves act as aerials and the changing currents radiate electromagnetic waves, further weakening the signal. This means that the signals need to be amplified at regular intervals.
- **low bandwidth** — the rate of transmission of information is limited.
- **noise** — unwanted signals (interference) are easily picked up. When the signal is amplified the noise is also amplified.

Coaxial cables
These are a development of the wire pair. The central wire, which acts as the transmitting wire, is sheathed by the outer conductor which is earthed and acts as the signal return path. It reduces the amount of noise picked up and reduces cross-linking. Coaxial cables have a larger bandwidth than copper wires, increasing the rate of transmission of information. They do not radiate electromagnetic waves to the same degree, which reduces attenuation. Security is slightly greater because they are somewhat more difficult to tap into.

Optic fibres
Optic fibres are increasingly replacing copper wires in telecommunications systems. They carry the information in the visible or near infrared region of the spectrum, at frequencies in the region of 10^{14} Hz. In theory, at such frequencies, hundreds of thousands of messages could be transmitted at the same time. In practice the number is limited by the frequency with which the lasers that pulse the light can be switched on and off. Nevertheless, optic fibres can work in the gigahertz wave region.

Advantages of optic fibres over copper-based wires include:
- much greater bandwidth, so can transmit information at a much faster rate

- less signal attenuation, so regeneration amplifying stations can be much further apart
- virtually impossible to tap, so much more secure
- negligible radiation of energy, so very little cross-linking
- much cheaper than their copper equivalent
- pick up much less noise, therefore there is a much clearer signal
- ideal for use with digital signals — the lasers producing the pulses can be switched on and off rapidly
- ideal for transcontinental communications — the cables are relatively cheap and of low weight so it is not over expensive to lay cables between and across continents; they can carry so many messages at the same time that only a few cables need to be laid across, even busy, routes

Radio waves and microwaves

Radio waves and microwaves are part of the electromagnetic spectrum, with frequencies ranging from about 30 kHz to 300 GHz. Although there is no fixed boundary between radio waves and microwaves, it is generally considered that microwaves have a frequency of 3 GHz or greater.

For convenience, radio waves are split into several further bands.

Table 23

Band	Frequencies	Wavelengths (in a vacuum)	Use
Long wave	30 kHz to 300 kHz	1 km to 10 km	Radio broadcast
Medium wave	300 kHz to 3 MHz	100 m to 1 km	Radio broadcast
Short wave	3 MHz to 30 MHz	10 m to 100 m	Radio broadcast
Very high frequency (VHF)	30 MHz to 300 MHz	1 m to 10 m	FM radio broadcast/ mobile phones
Ultra high frequency (UHF)	300 MHz to 3 GHz	10 cm to 1 m	Television broadcast/ mobile phones
Microwaves	3 GHz to 300 GHz	1 mm to 10 cm	Satellite broadcasts

Long-wave and, to a lesser extent, medium-wave radio waves have the advantage that they can diffract sufficiently to keep close to the Earth's surface and can also diffract round objects so that there are no 'shadows' in which reception is poor. Such radio waves are sometimes called **surface waves**.

Short-wave radio waves do not diffract sufficiently for this type of transmission. However, they do reflect from the layer of charged particles in the Earth's atmosphere (the ionosphere). It was in this way that Marconi made the first transatlantic radio broadcast. These are known as **sky waves**.

Waves in the VHF, the UHF and the microwave regions are of even shorter wavelength than short-wave radio waves and hence diffract even less. VHF and UHF waves penetrate the ionosphere and when used in communications are transmitted to satellites, which regenerate them and transmit them back to Earth. This type of wave is known as a **space wave**.

Table 24

Type of wave	Frequency/MHz	Wavelength (in a vacuum)/m	Range/km
Surface wave	<3	>100	1000
Sky wave	3 to 30	10 to 100	100*
Space wave	>30	<10	Line of sight*

*Both of these can be transmitted by reflection from the ionosphere or from regenerated waves from communications satellites, as appropriate. In practice, the use of the ionosphere as a reflecting layer is not used today because its lack of stability leads to inconsistent reception.

Bandwidth increases with increasing frequency, hence microwaves have a much greater bandwidth than even VHF and UHF radio waves. This means that information can be transmitted at a much faster rate using microwaves.

Communications satellites

Modern communications use satellites to relay messages. There are disadvantages. In particular, there is a delay between the sending of a signal and the recipient receiving the signal. In telephone conversations this leads to noticeable silences between replies. Nevertheless, they are invaluable for radio and television transmission, as well as long-distance telephone transmission where fibre-optic cables have not been laid.

The diagram below shows the principle of transmission by satellite link.

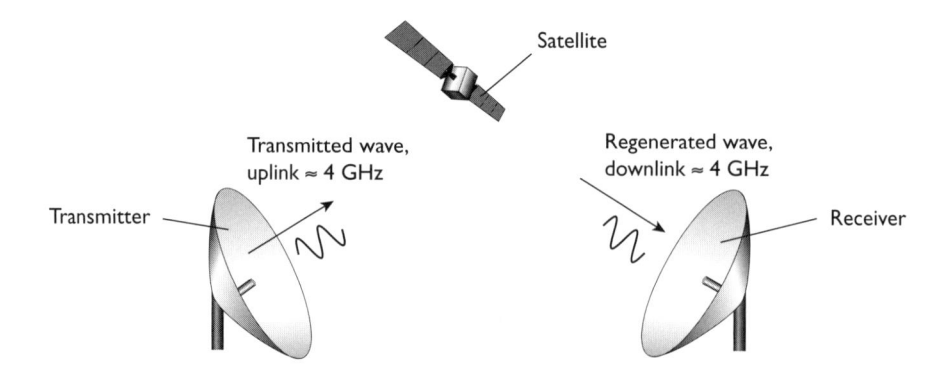

The signal reaching the satellite is only a tiny fraction of the transmitted signal. The satellite regenerates and amplifies the signal and retransmits it back to the Earth-based receivers at a different frequency. This change of frequency is required so that the powerful retransmitted signal does not swamp the much weaker signal from the Earth. The pathway from the Earth to the satellite is called the **uplink**; the path from the satellite to Earth is called the **downlink.**

Geostationary orbits
This is the type of orbit in which a satellite remains over the same point of the Earth.

> **Hint**
>
> You should refer back to geostationary orbits on page 141.

> **Worked example**
> When in a geostationary orbit, a satellite is about 3.6×10^7 m above the Earth. Estimate the time lapse between a signal being sent to the satellite and the receiver receiving the signal.
>
> **Answer**
> minimum distance the signal must travel = $2 \times 3.6 \times 10^7$ m.
>
> time = distance travelled/speed
>
> $t = s/v = 2 \times 3.6 \times 10^7 / (3.0 \times 10^8) = 0.24$ s

> **Note**
>
> This is the minimum delay. It is assumed that the path is straight up and down and that there is only one satellite link. Even in this case, it means that in a telephone conversation the gap between one person finishing a sentence and that person receiving the reply from the second person is twice this — nearly half a second. Consequently, where satellites are unable to do the job with a single link, optic fibres may be used in addition.

The major advantage of using satellites in geostationary orbits is that they are always over the same spot on the Earth. Therefore they do not need to be tracked and are in communication with the transmitter (and receiver) at all times.

Polar orbits

When in a polar orbit a satellite passes over the two poles. In general, polar orbits are much lower than geostationary orbits. They are about 1000 km above the Earth's surface with an orbital period of about 90 minutes. Therefore, they orbit the Earth about 16 times in a 24-hour period and cover every point on the Earth at some time each day. Polar orbital satellites are used for long-distance communication and have the advantage of considerably shorter delay times. However, they have to be tracked and the communication pathways have to be swapped from one satellite to another as they go beyond the horizon and as the Earth rotates below them. The major uses of polar orbital satellites are in studying the Earth's surface (e.g. to monitor crop growth or the melting of polar ice), weather forecasting and spying.

Attenuation

We have already met the idea of attenuation. This section looks at it in more detail, including the numerical calculation of attenuation.

Attenuation is the gradual decrease in power of a signal as it travels through space or a medium.

VII Gathering and communicating information

Attenuation can be very large and when two powers are compared they are measured on a logarithmic, rather than a linear, scale. The unit used to compare two powers P_1 and P_2 is the **bel**, where the number of **bels** is related to the powers by the equation:

number of bels = $\lg(P_1/P_2)$

It is more usual to use the decibel (1/10 of a bel). So,

number of decibels = $10\lg(P_1/P_2)$

The symbol for the bel is B and hence for decibels is dB.

> **Note**
>
> The abbreviation lg stands for log to the base 10. Most calculators show this as log or $\log_{10}$. Care must be taken not to confuse this with log to the base e, which is usually written as ln or $\log_e$.

> **Worked example**
> The attenuation of a signal along a wire is 30 dB. The initial signal has a power of 50 mW. Calculate the power of the signal after attenuation.
>
> *Answer*
> number of decibels = $10\lg(P_1/P_2)$
>
> $-30 = 10\lg(P_{out}/50)$
>
> $-3 = \lg(P_{out}/50)$
>
> $10^{-3} = (P_{out}/50)$
>
> $P_{out} = 50 \times 10^{-3} = 0.050$ mW

> **Note**
>
> The power output is less than the power input, so it is a negative number of decibels. With an amplifier, the output is (usually) greater than the input and the number of decibels is positive.

It may be useful to know the attenuation of a cable. Manufacturers give a typical figure of attenuation per unit length, measured in dB m^{-1}.

attenuation per unit length = attenuation/length of cable

Signal-to-noise ratio

As a signal travels along a cable the signal size decreases and the noise increases. It is unacceptable for the signal size to become so small that it is indistinguishable from the noise. The **minimum signal-to-noise ratio** gives a measure of the smallest signal amplitude that is acceptable.

Worked example

A cable has an attenuation of 4 dB km^{-1}. There is an input signal of power 0.75 W and the noise is 5×10^{-10} W. The minimum signal-to-noise ratio is 20 dB.

Calculate the maximum length of cable that can be used to carry this signal.

Answer

acceptable power-to-noise ratio = $10 \lg (P_s/P_{noise})$

$20 = 10 \lg (P_s/5 \times 10^{-10})$

$10^2 = P_s/5 \times 10^{-10}$

$P_s = 10^2 \times 5 \times 10^{-10} = 5 \times 10^{-8}$ W

maximum attenuation = $10 \lg (P_1/P_2) = 10 \lg (0.75/(5 \times 10^{-8})) = 72$

maximum length of cable = attenuation/attenuation per unit length

= 72/4 = 18 km

Note

It is also worth noting that the attenuation per unit length equals $(1/L)10 \log(P_1/P_2)$. This equation could be used in the last two parts of the example.

With an analogue signal, the noise is amplified as well as the signal. Hence the signal-to-noise ratio is unaltered, so this is the maximum distance that this signal could travel. With a digital signal, the signal can be regenerated as well as amplified.

Public switched telephone network (PSTN)

Landlines

To interconnect to different telephones a network is needed. The diagram explains the operation of a **public switched telephone network**.

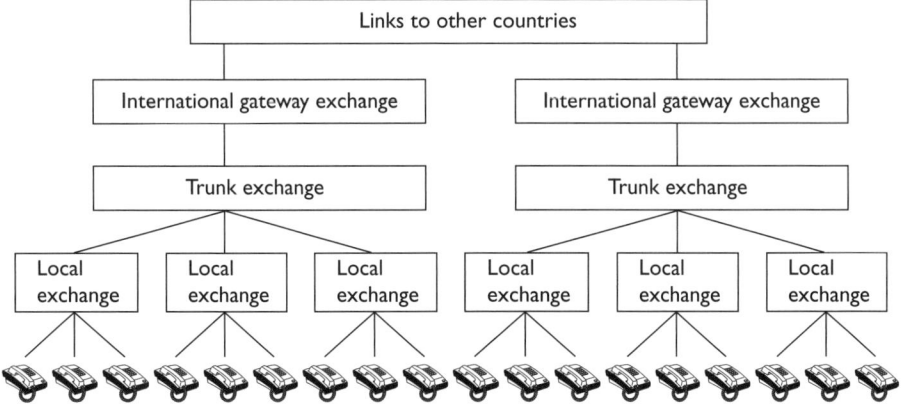

```
                    Links to other countries
           ┌─────────────────┴─────────────────┐
  International gateway exchange      International gateway exchange
           │                                   │
     Trunk exchange                      Trunk exchange
      ┌────┼────┐                         ┌────┼────┐
   Local Local Local                   Local Local Local
 exchange exchange exchange          exchange exchange exchange
```

Individual subscribers

Each subscriber is connected to a local exchange. When calling someone on the same local exchange, the subscriber will probably not need to dial an area code, but will be put through to the required number automatically. When calling someone on a different local exchange, the call goes through the trunk exchange and the subscriber has to include the area code. Likewise, when calling someone in a different country, that country's code will have to be included as well as the area code and the individual number.

Mobile phones
Communication lines

In America and some other countries mobile phones are known as cell phones. The name originates from the splitting of each area into smaller areas known as **cells**. Each cell is surrounded by six adjacent cells. Each of these adjacent cells uses a different set of frequencies, so there is no overlap. Because UHF and microwaves are used and the range of the signals is relatively small, more distant cells can share those frequencies. Each cell has a **base station** near its centre. The base station can communicate with all mobile phones in its cell, using several different frequencies. Just as the microwave signals sent to, and returned from, a satellite are of different frequencies, so are the signals sent from the base station to a mobile phone and the return signal from the phone. When a mobile phone is switched on it continually sends signals that several base stations pick up, and which identify that phone.

The base stations are all connected to the **cellular exchange**.

Table 25 follows the sequence of events when a handset is switched on.

Table 25

Action at handset	Action at base station(s)	Action at cellular exchange
Phone switched on and sends signals to identify itself	Receives signal and transfers to cellular exchange	Computers • select the base station with the strongest signal • allocate a carrier frequency for communication between base station and handset
Phone still switched on	Base stations continue to receive signals which they transfer to cellular exchange	Computers continue to monitor signals between handset and base stations
Phone still switched on moves from one cell to another	The signal at the original base station weakens and the signal becomes stronger at the new base station	Computers re-route the call through the new base station

The cellular exchange links one mobile phone directly to another via the base stations. However, there is a cable link from the cellular exchange to the PSTN so that mobile phones can communicate with landlines.

The handset

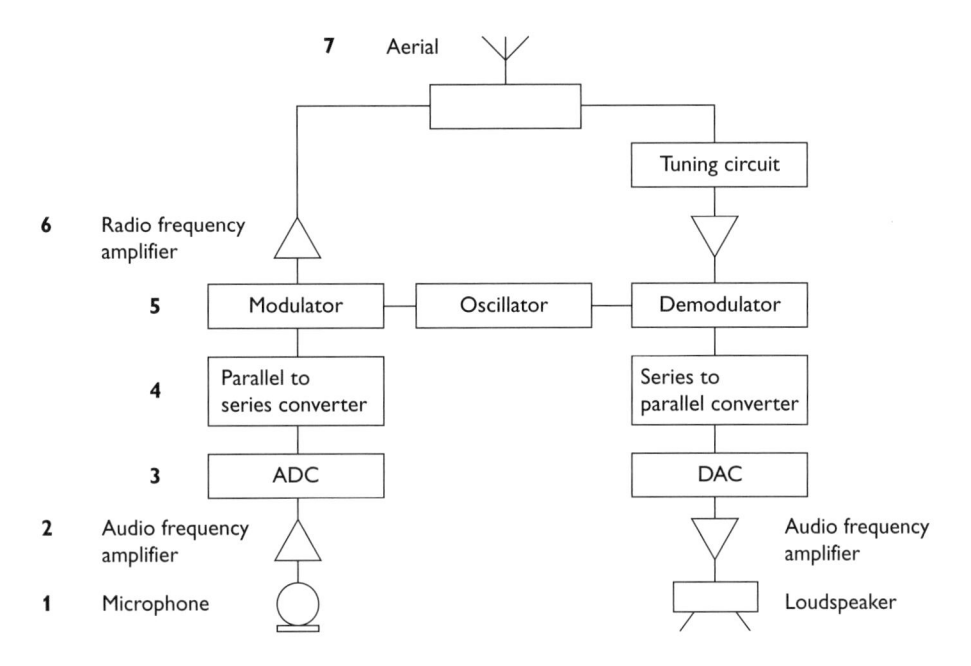

```
                           7    Aerial   Y
                        ┌─────────────────┐
                        │                 │
                        │         ┌───────┴───────┐
                        │         │ Tuning circuit │
                        │         └───────┬───────┘
  6  Radio frequency    △                 ▽
     amplifier
  5            ┌───────────┐  ┌──────────┐  ┌────────────┐
               │ Modulator │──│Oscillator│──│Demodulator │
               └─────┬─────┘  └──────────┘  └─────┬──────┘
               ┌─────┴──────┐                ┌────┴─────────┐
  4            │ Parallel to│                │  Series to   │
               │series conv.│                │parallel conv.│
               └─────┬──────┘                └────┬─────────┘
  3               ┌──┴──┐                      ┌──┴──┐
                  │ ADC │                      │ DAC │
                  └──┬──┘                      └──┬──┘
  2  Audio frequency △                            ▽  Audio frequency
     amplifier                                       amplifier
  1  Microphone   Q                               ⏵ Loudspeaker
```

Mobile phones work on an eight-bit digital system.

The diagram above is self explanatory — the list below describes the different stages.
1 The microphone converts the sound signal into an analogue electrical signal.
2 The analogue signal is amplified.
3 The ADC converts the analogue signal to digital.
4 The parallel-to-series converter takes each eight-bit binary number and emits it as a series of pulses (bits).
5 The carrier wave is modulated by the series of bits.
6 The modulated carrier wave is amplified.
7 The carrier wave is then switched to the aerial and sent to the base station.

The receiver acts in a similar but reverse manner to recreate an audio signal. The major difference is that the tuning circuit is used to pick up only the frequency that has been allocated to that particular handset.

A2 Experimental skills and investigations

Practical skills and investigations are examined on Paper 5, which is worth 30 marks. This is not a laboratory-based paper but, nevertheless, it tests the practical skills that you will have developed during the second year of your course. The table shows the breakdown of marks.

Skill	Total marks	Breakdown of marks	
Planning	15 marks	Defining the problem	3 marks
		Methods of data collection	5 marks
		Method of analysis	2 marks
		Safety considerations	1 mark
		Additional details	4 marks
Analysis, conclusions and evaluation	15 marks	Approach to data analysis	1 mark
		Table of results	2 marks
		Graph	3 marks
		Conclusion	4 marks
		Treatment of uncertainties	5 marks

The syllabus explains each of these skills in detail. It is important that you read the appropriate pages so that you know what each skill is, and what you will be tested on.

There is a great deal of information for you to take in and skills for you to develop. As with the skills that you developed in the AS year, these skills are learnt by doing practical work.

The examination questions

There are usually two questions on Paper 5, each worth 15 marks. Question 1 tests planning skills. Question 2 tests analysis, conclusions and evaluation. Read the questions carefully and make sure that you know what is being asked of you.

Question 1

This question asks you to plan an investigation. It is an open-ended question, which means that you need to think carefully about your answer before you start. There are various stages in planning an investigation.

Stage 1: define the problem
This requires you to look at the task that has been assigned and to identify the variables that impact on the problem:

- the independent variable (the variable that *you* control)
- the dependent variable (the variable that changes as a result of your changing the independent variable)
- any other variables that might affect your results, and which you need to control — generally by attempting to keep them constant

Stage 2: data collection
Once you have identified the variables, you have to decide on a method. You have to describe this method. You will almost certainly need to include a simple diagram to show the required apparatus. The description should also include how you are going to take measurements and how you intend to control any variables that might lead to the basic relationship between the independent and dependent variables being obscured. At this stage, you also need to think about any safety precautions you need to take.

Stage 3: analysis of results
The third stage in an investigation is the analysis of results. This will include:
- derived quantities that have to be calculated
- graphs that are to be plotted to identify the relationships between the variables

Tips

Before the exam: your practical course should include practice in designing and carrying out experiments. It is only by carrying out experiments that you will learn to look crically at the design and see the flaws in them.

In the exam: the examiner is highly unlikely to give you an experiment that you have met before, so do not be put off by something that seems unfamiliar. There will be prompts to guide you in answering the question. It is important that you look at these prompts and follow them carefully.

Question 2

This question takes an experiment that has been carried out and has had the results recorded for you. Your tasks are to analyse the results, including:
- calculating derived quantities and their uncertainties
- plotting suitable graphs to enable conclusions to be drawn
- evaluating the conclusions with regard to the calculated figures and the uncertainties

The syllabus requires that candidates should be able to:
- rearrange expressions into the forms $y = mx + c$, $y = ax^n$ and $y = ae^{kx}$
- plot a graph of y against x and use the graph to find the constants m and c in an equation of the form $y = mx + c$
- plot a graph of $\lg y$ against $\lg x$ and use the graph to find the constants a and n in an equation of the form $y = ax^n$
- plot a graph of $\ln y$ against x and use the graph to find the constants a and k in an equation of the form $y = ae^{kx}$

How to get high marks in Paper 5

Question 1

To demonstrate the stages in answering a question, it is useful to consider a particular question. Suppose the examiner asks you to investigate energy loss and its relationship with the thickness of a specific type of insulation.

Before starting you should have in your mind the sort of experiment that you would do to investigate these variables. There is no unique solution. One possibility is to put heated water in a beaker that has insulation wrapped around it and then to measure the rate of cooling of the water.

Stage 1

The independent variable is the thickness of insulation that is used.

The dependent variable is the energy lost from the container per unit time.

What other variables need to be controlled? Before you read any further, you should jot down some ideas.

Here are some thoughts that you might have considered:
- maintaining the same mass of water throughout the experiment
- the temperature of the surroundings should be kept constant
- the temperature fall during the test should be much smaller than the temperature difference between the container and the surroundings
- evaporation from the surface of any liquid used should be reduced to a minimum

Stage 2

The next task is to think about how you are going to carry out the experiment. Once you have a method in mind you need to:
- describe the method to be used to vary the independent variable
- describe how the independent variable is to be measured
- describe how the dependent variable is to be measured
- describe how other variables are to be controlled
- describe, with the aid of a clear labelled diagram, the arrangement of apparatus for the experiment and the procedures to be followed
- describe any safety precautions that you would take

In the experiment to investigate the energy loss through insulation, you may decide that the simplest way of varying the independent variable is to place the 'test beaker' inside a series of larger beakers and to fill the space between them with the insulating material. The thickness of the insulating material can then be calculated from the diameters of the different beakers and of the test beaker. These diameters could be measured using the internal jaws of a pair of vernier callipers. The energy loss could be measured by the temperature drop of the water in a specified time, or better, the time taken for a specified drop.

You should then describe how to ensure that other variables are controlled. You might use a top-pan balance to measure the mass of the test beaker and water between each set of readings. You could ensure that the water is at the same temperature each time by heating it in a constant temperature water bath (and then double checking the temperature before starting the stopwatch).

Think about safety. Does the water need to be near boiling? If you are heating it using a Bunsen burner then you should wear eye protection.

What extras might you include to ensure that your investigation is as accurate as possible? This tests your experience of doing practical work. Have you sufficient experience to see things that would improve the experiment?

Some of the ideas in the introduction to this part might be included. You might be able to think of some more:

- Choose a temperature drop that is much less than the difference between the starting temperature and room temperature.
- Stir the water in the bath so that it all reaches a uniform temperature.
- Make sure that the water is at the same starting temperature each time.
- Make sure that the room temperature is constant.
- Put a lid on the test beaker to prevent evaporation.
- Use a digital thermometer so that it is easy to spot when the temperature has fallen to a predetermined value.

Finally, a simple diagram of the apparatus is required. This will save a lot of description and can avoid ambiguities.

Hint

You might find it helpful to write out your description of the stages as a list of bullet points, rather than as continuous writing. Try it now with this example.

Stage 3

You will probably be told the type of relationship to expect. From this you should be able to decide what graph it would be sensible to plot. In this example, it might be suggested that the relationship between the variables is of the form:

$$(\Delta E/\Delta t) = ae^{-kx}$$

where $(\Delta E/\Delta t)$ is the rate of loss of energy, x is the thickness of the insulation and a and k are constants.

Taking logarithms of both sides of the equation gives:

$$\ln(\Delta E/\Delta t) = -kx + \ln a$$

Consequently, if you draw a graph of $\ln(\Delta E/\Delta t)$ against x it should be a straight line with a negative gradient. The gradient equals k and the intercept on the y-axis is equal to $\ln a$.

Question 2

How you tackle this part will depend on which relationship the question asks you to explore. The most likely relationships are of the form $y = ae^{kx}$ or $y = ax^n$.

Relationship $y = ae^{kx}$

To tackle this type of relationship you need to plot a graph of $\ln y$ against x.

> **Worked example**
> Consider the experiment described in Question 1. This table provides a possible set of results.
>
> | Thickness of insulation (x)/cm | Time taken for the temperature to fall 5.0°C/s | Rate of temperature fall $(\Delta T/\Delta t)/°C\,s^{-1}$ | $\ln((\Delta T/\Delta t)/°C\,s^{-1})$ |
> |---|---|---|---|
> | 2.0 | 110 | 0.0455 | −3.09 |
> | 3.1 | 127 | 0.0393 | −3.24 |
> | 4.3 | 149 | 0.0336 | −3.39 |
> | 5.1 | 165 | 0.0303 | −3.50 |
> | 6.4 | 195 | 0.0256 | −3.67 |
> | 7.2 | 216 | 0.0231 | −3.77 |
>
> Choose the points (1.0, −2.96) and (7.4, −3.80) to calculate the gradient:
>
> $$\text{gradient} = \frac{\Delta y}{\Delta x} = \frac{(-2.96 - (-3.80))}{(1.0 - 7.4)} = \frac{0.84}{-6.4} = -0.13$$
>
> Hence $k = -0.13\,\text{cm}^{-1}$
>
> To find a, choose a single point and use the generic equation for a straight line, $y = mx + c$. In this case, $y = \ln(\Delta T/\Delta t) + c$. So, choosing the point (1.0, −2.96):
>
> $-2.96 = -0.13 \times 1.0 + c$
>
> $c = -2.83$

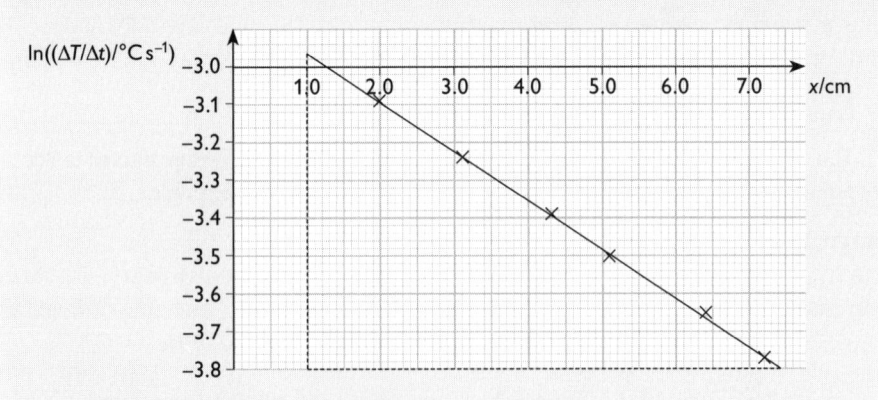

$c = \ln a$

Therefore:

$a = e^c = e^{-2.83} = 0.0590 \, °C\,s^{-1}$

Relationship $y = ax^n$

For this type of relationship you need to draw a graph of $\lg x$ against $\lg y$.

Worked example 1

An experiment is set up to investigate the diffraction of electrons by a carbon film. The diagram shows the experimental setup.

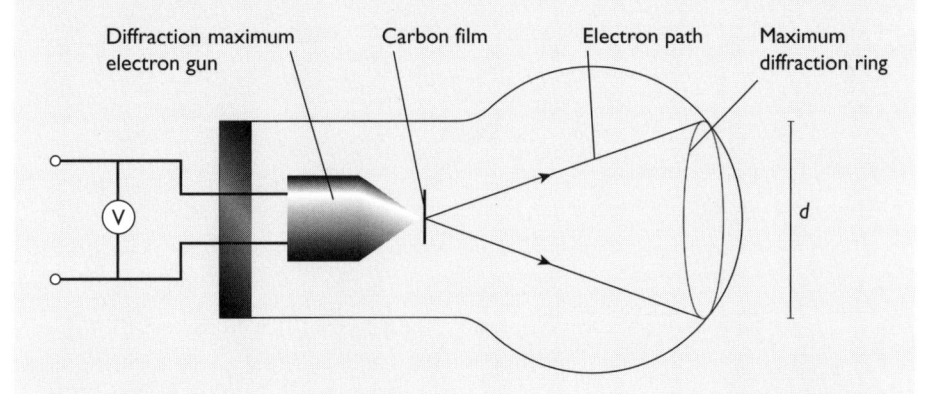

The diameter of the diffraction maximum ring was measured at different accelerating voltages.

The results are recorded in the table.

V/V × 10³	d/m × 10⁻²	lg (V/V)	lg (d/m)
2.0	9.8 ± 0.2	3.30	−1.01 ± 0.1
3.0	8.0 ± 0.2	3.48	−1.10 ± 0.1
4.0	6.9 ± 0.2	3.60	−1.16 ± 0.1
5.0	6.2 ± 0.2	3.70	−1.21 ± 0.2
6.0	5.6 ± 0.2	3.78	−1.25 ± 0.1

It is suggested that V and d are related by an equation of the form:

$d = kV^n$

where d is the diameter of the maximum ring, V is the accelerating voltage and k and n are constants.

Draw a graph to test the relationship between V and d. Include suitable error bars on your graph.

Draw the line of best fit through the points and also the worst acceptable line. Ensure that the lines are suitably labelled.

Answer
To solve this type of problem a graph of lg (V/V) against lg (d/m) is required.

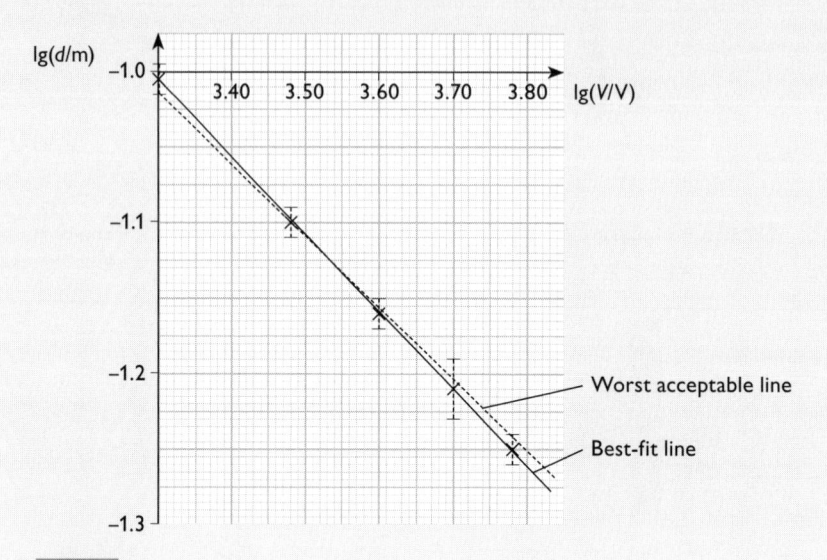

Note

The worst acceptable line is the line that has either the maximum or the minimum gradient and goes through all the error bars. In this example, the line of least gradient has been chosen.

You will observe that the uncertainties for d and for lg (d/m) are given. In an examination, you would be given the former but you would have to work out the latter. This is done by finding the value of lg (d/m) for the recorded value and for either the largest or smallest value in the range.

For example:

if $d/m = (8.0 \pm 0.2) \times 10^{-2}$

lg $(8.0 \times 10^{-2}) = -1.10$ and lg $(8.2 \times 10^{-2}) = -1.09$

uncertainty = $1.10 - 1.09 = 0.01$

It is important to realise that this process is required for every value of d.

The graph is a straight line, which confirms that the relationship of the form $d = kV^n$ is valid.

Worked example 2

(a) Calculate the gradient of the line. Include the uncertainty in your answer.

(b) Determine the intercept of the line of best fit. Include the uncertainty in your answer.

Answer

(a) For the gradient of the best-fit line, choose the points (3.30, −1.01) and (3.88, −1.30).

$$\text{gradient} = \frac{\Delta y}{\Delta x} = \frac{(-1.30 - (-1.01))}{(3.88 - 3.30)} = \frac{-0.29}{0.58} = -0.50$$

For the worst acceptable line, choose the points (3.30, −1.02) and (3.91, −1.30)

$$\text{gradient} = \frac{\Delta y}{\Delta x} = \frac{(-1.30 - (-1.02))}{(3.91 - 3.30)} = \frac{-0.28}{0.61} = -0.46$$

gradient = −0.50 ± 0.04

(b) To calculate the intercept, use the generic equation for a straight line:

$$y = mx + c$$

where, in this example, $y = \lg(d/m)$, m is the gradient = 0.50, $x = \lg(V/V)$ and c is the intercept = $\lg k$.

Choose the point (3.30, −1.01)

−1.01 = (−0.50 × 3.30) + c

c = 0.64

$k = 10^c = 10^{0.64} = 4.37$ (mV^{½})

To find the uncertainty in k, repeat the procedure using the worst acceptable line. Choose the point (3.30, −1.02), with a gradient 0.54.

−1.02 = (−0.54 × 3.30) + c

c = 0.762

$k = 10^c = 10^{0.762} = 5.78$ (mV^{½})

The uncertainty in k = ±(5.78 − 4.37) = ±1.41 (mV^{½})

From these answers the full expression for the relationship can be written down:

$d = 4.4V^{-0.5}$ which could be written:

$d = 4.4\sqrt{1/V}$

Note that the uncertainty in k makes it sensible to round to two significant figures.

A2 Questions & Answers

A2 Questions & Answers

Exemplar paper

This practice examination paper is similar to Paper 4. All the questions are based on the topic areas described in previous sections of this book.

You have 2 hours to complete the paper, which contains two sections. Section A covers work from syllabus sections I to VI and is worth 70 marks. Section B concentrates on Section VII, Gathering and communicating information, and is worth 30 marks. Therefore, there are 100 marks on the paper, so you can spend just over one minute per mark. If you aim for one minute per mark this will give you some leeway and perhaps time to check through your paper at the end.

See page 118 for advice on using this practice paper.

Section A

Question 1

(a) **Define gravitational potential at a point.** (1 mark)

(b) **A planet has a mass of 6.4×10^{23} kg and radius of 3.4×10^6 m.**

The planet may be considered to be isolated in space and to have its mass concentrated at its centre.

(i) Calculate the energy required to completely remove a spacecraft of mass 800 kg from the planet's surface to outer space. You may assume that the frictional forces are negligible. (3 marks)

(ii) A single short rocket burn was used for the spacecraft to escape from the surface of the planet to outer space. Calculate the minimum speed that the spacecraft would need to be given by the burn. (3 marks)

(c) **Calculate the gravitational field strength at the surface of the planet.** (2 marks)

Total: 9 marks

Candidate A
(a) Potential is the energy a mass has at a particular point in space ✗.

　The candidate has not picked up on the central point that potential refers to potential energy **per unit mass**. 0/1

(b) (i)　energy = −GMm/r ✓

　　　　= −6.67 × 10⁻¹¹ × 6.4 × 10²³ × 800/(3.4 × 10⁶) = −1.0 × 10¹⁰ J ✓ ✗

268

> This is quite a good answer. Unfortunately the candidate is a little confused with the signs. The potential energy at the surface is indeed negative, because it is an attractive field. However, to give the spacecraft more negative energy suggests that it is burrowing into the ground! This mark scheme is strict in that it insists on the presence of the minus sign. 2/3

(ii) kinetic energy = $½mv^2$ = 1.0×10^{10}

$0.5 \times 800 \times v^2 = 1.0 \times 10^{10}$

$v^2 = 1.0 \times 10^{10}/0.5 \times 800$

= $2.5 \times 10^7 \, m\,s^{-1}$ ✓ ✓ ✗

> A good start is made; everything is worked through correctly until at the end the square root of the value obtained for v^2 is not taken. 2/3

(c) $g = -GM/r^2 = 6.67 \times 10^{-11} \times 6.4 \times 10^{23}/(3.4 \times 10^6)^2 = 3.7 \, N\,kg^{-1}$ ✓ ✓

> Once more there is confusion over the minus sign, which suddenly disappears. However, a penalty for this was applied in part (b)(i), so there is no further penalty. 2/2

Candidate B

(a) Potential at a point is the work done in bringing unit mass from infinity to that point ✓.

> This is a perfect answer. It shows an understanding that potential is energy per unit mass and it uses the formal definition in terms of work. 1/1

(b) (i) potential energy of the body = $-GMm/r$

so the energy that must be given to the body =
$6.67 \times 10^{-11} \times 6.4 \times 10^{23} \times 800/(3.4 \times 10^6) = 1.0 \times 10^{10} \, J$ ✓ ✓ ✓

> This is an excellent answer. The explanation is clear and the working is laid out so that it is easy to understand. 3/3

(ii) kinetic energy = $½mv^2$ = the energy given for the spacecraft to escape from the planet

$0.5 \times 800 \times v^2 = 1.0 \times 10^{10}$

$v^2 = 1.0 \times 10^{10}/0.5 \times 800$

= $5000 \, m\,s^{-1}$ ✓ ✓ ✓

> This is another excellent clearly explained answer. 3/3

(c) $g = -GM/r^2 = -6.67 \times 10^{-11} \times 6.4 \times 10^{23}/(3.4 \times 10^6)^2 = -3.7 \, N\,kg^{-1}$

The minus sign shows that the acceleration is towards the centre of the planet ✓ ✓.

☺ This is another outstanding answer, with the added bonus of an explanation of the meaning of the minus sign. 2/2

Question 2

(a) Explain what is meant by internal energy (2 marks)

(b) 0.140 m³ of helium is contained in a cylinder by a frictionless piston. The piston is held in position so that the pressure of the helium is equal to atmospheric pressure and its temperature is 20°C.

(atmospheric pressure = 1.02×10^5 Pa)

  (i) Calculate the number of moles of helium in the container. (1 mark)

  (ii) Calculate the total kinetic energy of the helium atoms in the container. (3 marks)

(c) The temperature of the helium is gradually increased to 77°C and the helium expands against atmospheric pressure.

  (i) Calculate the volume of helium at 77°C. (1 mark)

  (ii) Calculate the total kinetic energy of the helium atoms at 77°C. (1 mark)

  (iii) Calculate the energy input to the helium. (2 marks)

Total: 10 marks

Candidate A

(a) Internal energy is the kinetic energy of a molecule in a body ✗ ✗.

☺ The candidate has some idea that internal energy is connected with the energy of the individual molecules but makes two serious errors. First, internal energy is not just the kinetic energy; it is the sum of the kinetic and potential energies. Second, considering a single molecule only is meaningless because an individual molecule is continually colliding and interacting with other molecules, so its kinetic and potential energies are continually changing. 0/2

(b) (i) $n = 1.02 \times 10^5 \times 0.14/(8.31 \times 20) = 85.9$ mol ✗

☺ This is a common error. The candidate has forgotten to convert degrees Celsius to kelvin. 0/1

(ii) $pV = \frac{1}{3} Nm \langle c^2 \rangle$, so $Nm \langle c^2 \rangle = 3pV = 3 \times 1.02 \times 10^5 \times 0.14 = 42\,840$ ✓

$E_k = \frac{1}{2} mv^2 = 42\,840/2 = 21\,420$ J ✓ ✓

☺ This is done well. However, the lack of explanation means that it would have been difficult to award part marks had there been an arithmetical error. There is no penalty for the extra significant figure. 3/3

(c) (i) $p_1V_1/T_1 = p_2V_2/T_2$

0.14/20 = V_2/77

V_2 = 0.539 m³ ✓ (e.c.f.)

📝 This is an acceptable method of finding the new volume. However, the mistake of not converting to kelvin is repeated. It has not been penalised a second time, hence the error carried forward (e.c.f.). 1/1

(ii) $pV = \frac{1}{3}Nm\langle c^2 \rangle$, so $Nm\langle c^2 \rangle = 3pV$ = 3 × 1.02 × 10⁵ × 0.539 = 164 934 J

E_k = ½ × 164 934 = 82 467 J ✓

📝 This is calculated correctly. 1/1

(iii) energy input = 82 467 − 42 840 = 39 627 J[c] ✗

📝 The candidate has simply found the differences in the kinetic energies, and has not recognised that the gas does work in expanding and therefore loses potential energy. 0/2

Candidate B

(a) Internal energy is the sum of the random kinetic and potential energies of the particles in the body.

📝 This is an excellent answer. The candidate has a clear idea of the concept. 2/2

(b) (i) $pV = nRT$

1.02 × 10⁵ × 0.140 = n × 8.31 × 293

n = 5.86 ✓

📝 The calculation is correct. 1/1

(ii) E_k for one atom = $\frac{3}{2}kT$ = $\frac{3}{2}$ × 1.38 × 10⁻²³ × 293 = 6.07 × 10⁻²¹ J ✓

total kinetic energy = energy of 1 atom × number of atoms

total E_k = 6.07 × 10⁻²¹ × 5.86 × 6.02 × 10²³ = 2.14 × 10⁴ J ✓ ✓

📝 This is done well. The candidate has used a different method from that used by Candidate A. The use of the equation $E_k = \frac{3}{2}kT$ is a good way of solving the problem. 3/3

(c) (i) $pV = nRT$

V = 5.86 × 8.31 × 350/(1.02 × 10⁵) = 0.167 m³ ✓

📝 The candidate has successfully applied the ideal gas equation. 1/1

(ii) temperature is proportional to E_k.

ratio of temperatures = ratio of E_k

350/293 = new E_k/2.14 × 10⁴

A2 Questions & Answers

new $E_k = 2.56 \times 10^4$ J ✓

☕ This is a neat, if slightly risky, way of doing this calculation. It might be more orthodox to go through the $E_k = {}^3/_2 kTE_k = {}^3/_2 kT$ calculation again. 1/1

(iii) $\Delta U = \Delta Q + \Delta W = \Delta Q + p\Delta V$

$\Delta Q = \Delta U - p\Delta V = (2.56 - 2.14) \times 10^4 - ((0.140 - 0.167) \times 1.02 \times 10^5) = 4200 + 2750$
$\approx 6.95 \times 10^4$ J ✓ ✓

☕ This is an outstanding answer. The candidate clearly understands the physics and has worked through the problem sensibly. Notice that ΔW is negative because the gas does work on the atmosphere, rather than having work done on itself. 2/2

Question 3

The pendulum bob on a large clock has a mass of 0.75 kg and oscillates with simple harmonic motion. It has a period of 2.0 s and an amplitude of 12 cm.

(a) Calculate the maximum restoring force on the pendulum bob. (2 marks)

(b) (i) Calculate the maximum kinetic energy of the bob. (3 marks)

(ii) State the maximum potential energy of the bob. (1 mark)

(c) If the clock is not wound up the oscillation of the pendulum is lightly damped.

(i) Explain what is meant by 'lightly damped'. (1 mark)

(ii) Draw a graph on the grid to show the oscillation of this lightly damped oscillation. (2 marks)

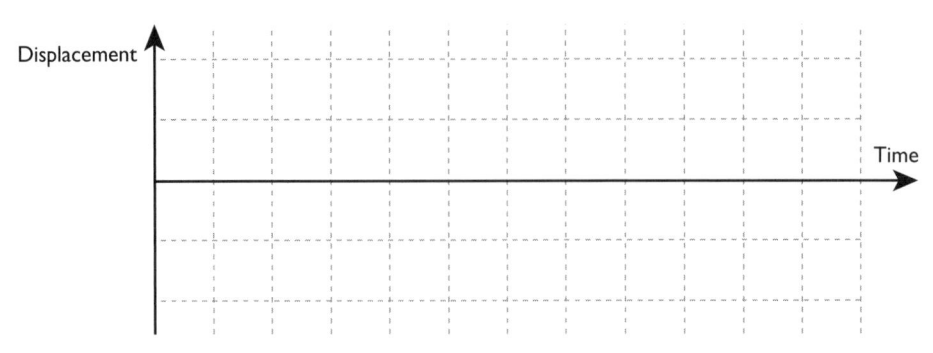

Displacement ↑ → Time

Total: 9 marks

Candidate A
(a) $F = m\omega^2 r$

$F = 0.75 \times \pi^2 \times 12 = 89$ N ✓ ✗

> The working is reasonably clear and gets the correct answer, although it is not quite clear where the π^2 comes from (it is ω^2). Unfortunately, in the equation the amplitude is not converted into metres. **1/2**

(b) (i) maximum speed = $\omega r = \pi \times 12 = 38\,\text{cm}\,\text{s}^{-1}$ ✓

$E_k = \frac{1}{2}mv^2 = 0.5 \times 0.75 \times 38^2 = 542\,\text{J}$ ✓ ✓

(ii) $mgh = ?$ ✗

> This is a repeated error, only this time not converting $\text{cm}\,\text{s}^{-1}$ into $\text{m}\,\text{s}^{-1}$. This is not penalised a second time, although the candidate should realise that the answer obtained is far too large. The next part is not answered because the candidate does not understand that the kinetic energy and the potential energy add to give the total energy, which remains constant. **2/4**

(c) Its amplitude is decreasing.

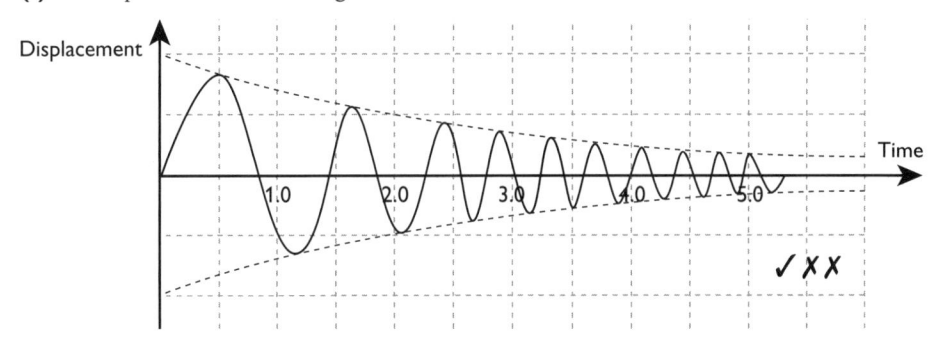

> There is some understanding of light damping but the graph is poor. Although the envelope is correct, the question asks for a graph of this particular motion, which indicates some values are needed — in this case the period. The candidate does not show the period as 2s but reduces the period as the amplitude decreases. **1/3**

Candidate B
(a) $F = -m\omega^2 r$

$\omega = 2\pi/T = 2\pi/2 = \pi$

$F = -0.75 \times \pi^2 \times 0.12 = -0.89\,\text{N}$ ✓ ✓

> There is clear working here to arrive at the correct answer, with the understanding that the maximum restoring force is at maximum displacement. The inclusion of the minus sign, while not an absolute requirement, shows an appreciation that the force is a restoring force. **2/2**

(b) (i) maximum speed = $\omega r = \pi \times 0.12 = 0.38\,\text{m}\,\text{s}^{-1}$ ✓

$E_k = \frac{1}{2}mv^2 = 0.5 \times 0.75 \times 0.38^2 = 0.054\,\text{J}$ ✓ ✓

(ii) $0.054\,\text{J}$ ✓

> Once more there is clear working leading to the correct answer, and then a clear understanding that the kinetic energy is converted to potential energy.
> 4/4

(c) It is having to do work against something like friction, so it gradually loses energy and the amplitude decreases slowly ✓.

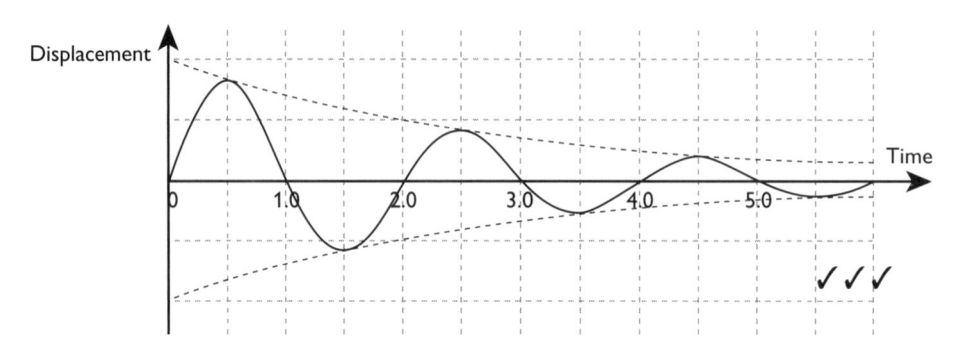

> The answer shows a clear understanding of damping and the diagram confirms an understanding of light damping. The correct (and constant) period is pleasing. 3/3

Question 4

The spherical dome on a Van de Graaff generator is placed near an earthed metal plate.

Consider the dome as an isolated sphere with all its charge concentrated at its centre.

(a) The dome has a diameter of 50 cm and the potential at its surface is 65 kV.

   (i) Calculate the charge on the dome. (2 marks)

   (ii) Calculate the capacitance of the dome. (1 mark)

The metal plate is moved slowly towards the dome and it partially discharges through the plate, leaving the dome with a potential of 12 kV.

(b) Calculate the energy that is dissipated during the discharge. (4 marks)

Total: 7 marks

Candidate A

(a) (i) $V = \dfrac{1}{4\pi\varepsilon_0} \dfrac{Q}{r}$

$65 \times 10^3 = \dfrac{1}{4\pi \times 8.85 \times 10^{-12}} \dfrac{Q}{0.5}$

$Q = 3.6 \times 10^{-6}$ C ✓ ✗

(ii) $C = Q/V = 3.60 \times 10^{-6}/(65 \times 10^3) = 5.5 \times 10^{-11}$ F ✓ (e.c.f.)

📝 This is a reasonable effort. It is correct apart from the diameter (0.5 m), rather than the radius (0.25 m), of the dome being used. 2/3

(b) change in voltage = 55 kV, therefore

change in energy = ½QV = 0.5 × 3.60 ×10⁻⁶ × 55 × 10³ = 9.9 × 10⁻² J ✓ ✗ ✗ ✗

📝 The candidate is taking the energy to vary linearly with the voltage — it varies with voltage squared. The easiest way of calculating the discharge energy is to find both the energy before and the energy after discharge. Then find the difference between them. A compensation mark is given for the use of the formula ½QV. 1/4

Candidate B

(a) (i) $V = \dfrac{1}{4\pi\varepsilon_0} \dfrac{Q}{r}$

$65 \times 10^3 = \dfrac{1}{4\pi \times 8.85 \times 10^{-12}} \dfrac{Q}{0.25}$

$Q = 1.80 \times 10^{-6}$ C ✓ ✓

(ii) $C = Q/V = 1.80 \times 10^{-6}/(65 \times 10^3) = 2.8 \times 10^{-11}$ F ✓

📝 These are good clear calculations, resulting in the correct answers. 3/3

(b) energy before discharge = ½QV = 0.5 × 1.80 × 10⁻⁶ × 65 × 10³ = 5.9 × 10⁻² J ✓ ✓

energy after discharge = ½CV² = 0.5 × 2.8 × 10⁻¹¹ × (12 × 10³)² = 2.0 × 10⁻³ J ✓

energy dissipated = (5.9 − 0.2) × 10⁻² J = 5.7 × 10⁻² J ✓

📝 The problem is tackled in a logical manner and the working is once more easy to follow. 4/4

Question 5

(a) State Faraday's law of electromagnetic induction. (2 marks)

(b) An aeroplane is flying at a steady altitude in a direction perpendicular to the Earth's magnetic axis. The Earth's magnetic field has a flux density of 34 µT and it makes an angle of 60° with the Earth's surface. The wingspan of the aeroplane is 42 m and it is travelling at a speed of 180 m s⁻¹.

  (i) Calculate the e.m.f. induced across the wings of the aeroplane. (3 marks)

  (ii) Explain why this e.m.f. could not drive a current through a conductor connected across the wingtips of the aeroplane. (1 mark)

(iii) State and explain the effect on the e.m.f. across the wingtips if the aeroplane were travelling parallel to the Earth's magnetic axis. *(2 marks)*

Total: 8 marks

Candidate A
(a) An induced e.m.f. is equal to the rate of cutting magnetic flux ✓ ✓.

> The candidate has some concept of Faraday's law. The idea of cutting flux is a useful model, although it could be argued that it excludes a change in flux density. Another fault is that it does not include the idea of flux linkage, which includes the number of turns in a coil. This definition only really caters for a single wire in the field. Nevertheless, the examiner has given the candidate the benefit of the doubt. 2/2

(b) (i) $E = BA\sin\theta$

$E = 34 \times 10^{-6} \times 180 \times 42 \times \sin 30 = 0.13\,\text{V}$ ✓ ✗ ✗

> This is not explained well. The candidate has worked out the area swept out by multiplying the speed by the wingspan but has used the incorrect angle. It is the vertical component that induces the e.m.f. across the wings, not the horizontal component. 1/3

(ii) The voltage is too small to drive a meaningful current ✗.

> The candidate has not spotted the fact that the conductor would have to move along with the aeroplane. 0/1

(iii) There would be no voltage as the aeroplane would be flying parallel to the flux ✗ ✗.

> It is the vertical component of the field that induces an e.m.f. across the wings. The horizontal component would induce an e.m.f. between the top and the bottom of the aeroplane. 0/2

Candidate B
(a) The magnitude of an induced e.m.f. is equal to the rate of change of magnetic flux linkage ✓ ✓.

> This is a textbook definition that includes all the relevant points. 2/2

(b) (i) $E = -\Delta\phi/\Delta t = (\Delta A/\Delta t)B\sin\theta$

area swept out per unit time = $Lv = 42 \times 180 = 7560\,\text{m}^2\text{s}^{-1}$ ✓

vertical component of the flux induces this e.m.f. = $B\sin 60$ ✓

$E = 7560 \times 34 \times 10^{-6} \times \sin 60 = 0.22\,\text{V}$ ✓

> This is done well. The candidate shows each stage of the calculation clearly. 3/3

(ii) The conductor would also travel through the magnetic field and therefore have the same e.m.f. induced across it ✓.

International AS and A Level Physics Revision Guide

> This is a good answer. 1/1

(iii) There would be no change ✓. It is the vertical component that induces the field across the wings ✓.

> The candidate has spotted the important factor and explained it concisely. 2/2

Question 6

(a) Electrons are accelerated through a potential difference of 4.8 keV.

   Calculate the velocity of the electrons   (3 marks)

In a different experiment, electrons travelling at a speed of $2.8 \times 10^7 \,\mathrm{m\,s^{-1}}$ enter a uniform magnetic field perpendicularly to the field. The magnetic field has a flux density 4.0 mT.

(b) (i) Explain why the electrons travel in a circular path in the magnetic field.   (2 marks)

   (ii) Calculate the magnitude of the force on the electrons due to the magnetic field.   (2 marks)

   (iii) Calculate the radius of the circular path of the electrons.   (2 marks)

(c) As the electrons travel through the field they gradually lose energy.

   State and explain the effect of this on the radius of the path.   (2 marks)

   Total: 11 marks

Candidate A

(a) energy = eV = $1.6 \times 10^{-19} \times 4.8 = 7.68 \times 10^{-19}$ J ✗

$7.68 \times 10^{-19} = 0.5 \times 9.1 \times 10^{-31} \times v^2$

$v = \sqrt{(7.68 \times 10^{-16})/(0.5 \times 9.1 \times 10^{-31})} = 1.3 \times 10^6 \,\mathrm{m\,s^{-1}}$ ✓✓ (e.c.f.)

> This is a good effort; although the candidate forgot to change kilovolts to volts. The work is set out reasonably well, although it would be improved by including the expression for kinetic energy. Nevertheless, it is easy to spot the mistake, so only one mark is lost. 2/3

(b) (i) The force is at right angles to the velocity ✓ so the motion is circular.

> The candidate scores the first mark for recognising that the force is at right angles to the velocity but does not develop the argument. 1/2

(ii) $F = Bqv = 4.0 \times 10^{-3} \times 1.6 \times 10^{-19} \times 2.8 \times 10^7 = 1.79 \times 10^{-14}$ N ✓✓

> This part is done well and scores both marks. 2/2

(iii) $F = \tfrac{1}{2}mv^2/r$ ✗

$1.79 \times 10^{-14} = \frac{1}{2} \times 9.1 \times 10^{-31} \times (2.8 \times 10^7)^2 / r$

$r = \frac{1}{2} \times 9.1 \times 10^{-31} \times (2.8 \times 10^7)^2 / (1.79 \times 10^{-14}) = 1.99 \times 10^{-2}$ m ✗

> The formula is wrong. The candidate has added in a '½', as if it were kinetic energy. This is a serious error showing a failure to understand the physics, so both marks are lost. 0/2

(c) The radius is reduced ✓ because it deceases when the velocity decreases ✗.

> The candidate recognises that the radius decreases — perhaps the result is remembered from the experiment being carried out during the course. However, the reasoning simply repeats what has been said already. 1/2

Candidate B

(a) energy of the electrons $= eV = 1.6 \times 10^{-19} \times 4.8 \times 10^3 = 7.68 \times 10^{-16}$ J

$E_k = \frac{1}{2} m v^2$

$7.68 \times 10^{-16} = 0.5 \times 9.1 \times 10^{-31} \times v^2$

$v = \sqrt{(7.68 \times 10^{-16})/(0.5 \times 9.1 \times 10^{-31})} = 4.1 \times 10^7$ m s^{-1} ✓✓✓

> This is a clear, well set out and correct calculation. 3/3

(b) (i) The force is at right angles to the velocity of the electrons, so there is no change in the magnitude of the velocity. As the direction of the velocity changes, so does the force direction, always remaining at right angles to the velocity. The magnitude of the velocity remains constant ✓✓.

> This excellent description takes the argument further than Candidate A, explaining how the direction of the force changes continuously as the velocity direction changes. 2/2

(ii) $F = Bev = 4.0 \times 10^{-3} \times 1.6 \times 10^{-19} \times 2.8 \times 10^7 = 1.79 \times 10^{-14}$ N ✓✓

> This shows good, clear use of the equation for the force on a charged particle in a magnetic field. The candidate uses e for the charge, because it is the charge on an electron. 2/2

(iii) $F = mv^2/r$

$1.79 \times 10^{-14} = 9.1 \times 10^{-31} \times (2.8 \times 10^7)^2 / r$ ✓

$r = 9.1 \times 10^{-31} \times (2.8 \times 10^7)^2 / (1.79 \times 10^{-14}) = 3.98 \times 10^{-2}$ m ✓

> This is clear use of the equation for circular motion. 2/2

(c) The radius is reduced because the velocity falls. The centripetal force is equal to $Bqv = mv^2/r$, thus $r = mv/Bq$. B, q and m are unchanged therefore r is proportional to v.

> This is a well reasoned argument. 2/2

Question 7

The Planck constant links both the wave–particle duality of matter and of electromagnetic radiation.

(a) State the equations that show this duality. (2 marks)

(b) Describe the photoelectric effect and explain why it gives evidence for the wave–particle duality of electromagnetic radiation. (4 marks)

(c) (i) Aluminium has a work function energy of 6.52×10^{-19} J. Calculate the maximum kinetic energy with which an electron can be emitted from this metal when electromagnetic radiation of wavelength, 1.80×10^{-7} m falls on its surface. (3 marks)

(ii) State which part of the electromagnet spectrum this radiation is in. (1 mark)

Total: 10 marks

Candidate A

(a) $E = hc/\lambda$

$\lambda = h/p$ ✓ ✗

> The equations are correct. However, the candidate has not explained what the symbols mean, so a mark is lost. 1/2

(b) Photoelectric effect is the emission of electrons from a metal surface when light falls on it ✗. There is a minimum frequency at which electrons are emitted. If light were waves there would be no minimum ✓, so light comes in packets of energy called photons ✗ ✗.

> This is not a good description. The photoelectric effect is not just about visible light but about all electromagnetic radiation, so the first mark is lost. The candidate has some idea of the reasoning but does not explore it very deeply. The comment regarding photons is correct but not really relevant to the argument and requires more development. 1/4

(c) (i) $E = hc/\lambda = 6.63 \times 10^{-34} \times 3 \times 10^8 / (1.8 \times 10^{-7}) = 1.105 \times 10^{-18}$ ✓

energy of the photon = E_k + work function = $1.105 \times 10^{-18} + 6.52 \times 10^{-19}$
$= 1.76 \times 10^{-19}$ J ✗ ✗

> The candidate starts off well, calculating the energy of the photon and then correctly writing down the equation. However, not enough care has been taken in rearranging the equation. (1/3)

(ii) X-rays ✗

> You need to know the rough boundaries of the parts of the electromagnetic spectrum. This radiation is well into the ultraviolet. 0/1

Candidate B

(a) $E = hf$, where h = the Planck constant, E = the photon energy and f = the frequency of the radiation ✓

$\lambda = h/p$, where λ is the electron wavelength, h is the Planck constant and p is the momentum of the electron ✓

☺ Both are correct. The equation for electromagnetic radiation can be used with either frequency or wavelength. The symbols used are explained clearly. 2/2

(b) Photoelectric effect is the emission of electrons from a metal surface when electromagnetic radiation falls on it ✓. If electromagnetic radiation were purely wave-like in nature then radiation of all frequencies would cause the effect ✓. However, there is a minimum frequency radiation, which is different for all metals, below which no electrons are emitted ✓. This is called the threshold frequency. The electrons are emitted immediately radiation above the threshold frequency falls on the metal. There is no wait for the continuous wave to supply enough energy ✓.

☺ This is an excellent description. The effect is described correctly and the points are made clearly. The answer is much deeper than that of Candidate A. The immediate emission of electrons is explored and there is discussion of the why the pure wave model does not work. 4/4

(c) (i) $E = hc/\lambda = 6.63 \times 10^{-34} \times 3 \times 10^8/(1.8 \times 10^{-7}) = 1.105 \times 10^{18}$ J ✓

E_k = energy of the photon − work function = $(11.05 − 6.52) \times 10^{-19}$
= 4.53×10^{-19} J ✓ ✓

☺ This is worked through well. 3/3

(ii) Visible light ✗

☺ This is a rare error. The range of wavelength of visible radiation is from about 4×10^{-7} m (violet) to 7×10^{-7} m (red). 0/1

Question 8

(a) For nuclear fusion to occur, temperatures in the region of 1 million degrees kelvin are required.

Explain why such a high temperature is needed. (2 marks)

(b) One form of nuclear fusion in stars is known as the proton–proton chain. In this chain a total of six protons combine to form an alpha particle. In addition to the formation of the alpha particle, two protons and two positrons are released.

Calculate the energy, in joules, released in the proton–proton chain. (4 marks)

(mass of an alpha particle = 6.64424 × 10⁻²⁷ kg, mass of a proton = 1.67261 × 10⁻²⁷ kg, mass of a positron = 9.1 × 10⁻³¹ kg)

Total: 6 marks

Candidate A

(a) The very large electrostatic repulsion between nuclei as they approach to within fusion distances means they must have a very high speed ✓.

> The candidate has the basic idea but does not go on to explain the link between mean velocity of the particles and temperature. 1/2

(b) mass of six protons = $6 \times 1.67261 \times 10^{-27} = 10.03566 \times 10^{-27}$ kg ✓

mass of products = $6.64424 \times 10^{-27} + (2 \times 9.1 \times 10^{-31}) = 6.64606 \times 10^{-27}$ kg ✗

mass lost = $(10.03566 - 6.64606) \times 10^{-27} = 3.3896 \times 10^{-27}$ kg ✓ (e.c.f.)

$E = mc^2 = 3.3896 \times 10^{-27} \times (3.0 \times 10^8)^2 = 3.05 \times 10^{-10}$ J ✓

> This is a good effort. Unfortunately, the two protons released in the process have been missed. Apart from that omission, the answer shows a clear understanding of the physics. Error carried forward (e.c.f.) means that only 1 mark is lost. 3/4

Candidate B

(a) The very large electrostatic repulsion between nuclei as they approach to within fusion distances means they must travel at very high speed ✓. Temperature is proportional to the mean square speed ✓.

> This excellent answer takes the argument a step further than Candidate A. 2/2

(b) net number of protons = 4

mass of four protons = 4 × initial mass of protons = $4 \times 1.67261 \times 10^{-27}$
= 6.69044×10^{-27} kg ✓

mass of products = $6.64424 \times 10^{-27} + (2 \times 9.1 \times 10^{-31}) = 6.64606 \times 10^{-27}$ kg ✓

mass lost = $(6.69044 - 6.64606) \times 10^{-27} = 0.04438 \times 10^{-27}$ kg ✓

$E = mc^2 = 0.04438 \times 10^{-27} \times (3.0 \times 10^8)^2 = 3.99 \times 10^{-12}$ J ✓

> This is worked through in a logical way, with all the steps shown clearly. 4/4

A2 Questions & Answers

Section B

Question 9

(a) (i) Explain what is meant by negative feedback in an amplifier. (2 marks)

    (ii) State two advantages of having negative feedback with an operating amplifier. (2 marks)

(b) The circuit diagram shows an operational amplifier as an inverting amplifier.

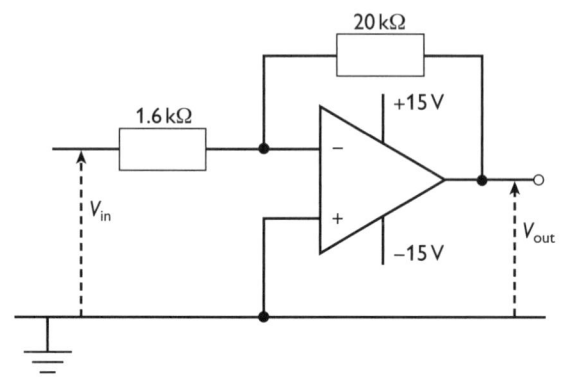

Calculate the voltage output when the input voltage is:

(i) 0.60 V (2 marks)

(ii) 1.8 V (1 mark)

(c) An operational amplifier is used as a comparator to switch a 2 kW heater on when the temperature falls below a specified temperature.

Explain, with the aid of a diagram, how the output from the operational amplifier could be used to switch on the heater. You do not need to draw the op-amp circuit. (4 marks)

Total: 11 marks

Candidate A

(a) (i) Some of the output is fed back to the input ✗ ✗.

> 🅔 The candidate has some idea of feedback but does not refer to its negative aspect. The use of the term 'fed back' to describe feedback is not wise, as it paraphrases the question. The benefit of the doubt is not given. 0/2

    (ii) There is less distortion due to the op-amp saturating ✓. There is a wider frequency range ✗.

The first point is correct but the second part is unclear. Does it mean that there is a wider bandwidth or that there is more consistent amplification over a wider frequency range? 1/2

(b) (i) gain = R_f/R_{in} = 20/1.6 = 12.5 ✗

$V_{out} = AV_{in}$ = 12.5 × 0.6 = 7.5 V ✓ (e.c.f.)

(ii) $V_{out} = AV_{in}$ = 12.5 × 1.8 = 22.5 V ✗

 The candidate misses the idea of the inverting amplifier and gives the gain as a positive number. This feeds through the answers. The fact that the amplifier saturates when the input voltage is 1.8 V is not recognised. 1/3

(c)

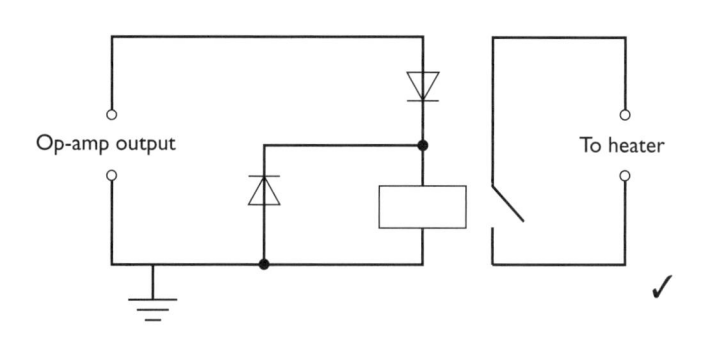

The heater would require a larger power input than the op-amp can supply ✓. A relay is used to switch the heater which is run from a high-power circuit. The diodes are used to protect the op-amp ✓.

 This is quite good. A little more detail on what the diodes protect the operational amplifier from would make the answer excellent. 3/4

Candidate B

(a) (i) Some of the output which is out of phase with the input is returned to the input ✓ ✓.

 This is a good, succinct answer that includes the negative nature of the feedback. 2/2

(ii) There is a more consistent gain over a wider range of frequencies and there is less distortion due to the op-amp saturating ✓ ✓.

 This is a fully correct answer that includes both the relevant points. 2/2

(b) (i) gain = $-R_f/R_{in}$ = −20/1.6 = −12.5 ✓

$V_{out} = AV_{in}$ = −12.5 × 0.6 = −7.5 V ✓

(ii) $V_{out} = AV_{in}$ = −12.5 × 1.8 = −22.5 V but the supply voltage is ±15 V, so the op-amp saturates and the output is −15 V ✓

 This fully correct answer scores full marks. 3/3

(c)

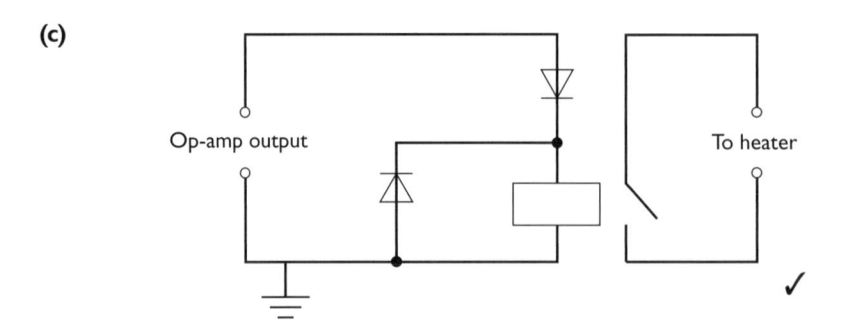

[Circuit diagram showing Op-amp output connected to a relay circuit with diodes, leading To heater ✓]

The heater would require a larger power input than the op-amp can supply ✓. A relay is used to switch the heater which is run from a separate high-power circuit. The diodes are used to protect the op-amp ✓ because a large back e.m.f. is induced across the ends of the relay coil when it opens ✓.

🄔 This full answer explains why the diodes are necessary. The candidate might have gone on to explain the logic of how the diodes protect the op-amp. A mark has been awarded for the correct diagram. However, there are only 4 marks for this part-question and three further relevant points have been made already. 4/4

Question 10

(a) Describe the differences in the production of a CT-scan image with the production of a traditional X-ray image. (6 marks)

(b) (i) Describe the advantages of an MRI scan compared with a CT scan. (2 marks)

(ii) Explain why it is sometimes useful to produce a combined image from a CT scan and an MRI scan. (2 marks)

Total: 10 marks

Candidate A

(a) An X-ray image is a two-dimensional image formed by shining a beam of X-rays onto a photographic plate. CT scanning produces a three-dimensional image ✓ by making images of many slices ✓.

🄔 This is a reasonable answer, which, with a little more detail could have scored more marks. The candidate understands the process of taking the X-ray image but should describe it as a 'shadow image', rather than as a two-dimensional image. There is no description of how the X-ray image is formed. The description of the CT image lacks detail and so does not gain full credit. 2/6

(b) (i) CT scans do not give as good contrast with fleshy tissues as MRI scans ✓. MRI scans cannot be used with people who have metallic replacement parts in their bodies ✗.

> The question has been answered as a disadvantage of CT scans but the point made is relevant and well worth a mark. The comment regarding the metallic replacement parts is irrelevant — if anything, this is a disadvantage of the MRI-scanning procedure, not an advantage. 1/2

(ii) The combination gives doctors the best results from both methods and gives them more information ✗ ✗.

> The candidate has said nothing relevant here. What are the best results from each? 0/2

Candidate B

(a) An X-ray image is a shadow image ✓ caused by the differing absorption ✓ of the different body tissues as a single beam of X-rays is passed through the patient. In a CT scan the patient is in the centre of a ring and the X-ray source rotates around the patient taking a series of images through a slice ✓. The information gained is sent to a computer which builds up a picture of the slice ✓. The patient is moved slowly through the machine so that this is repeated to make images of many slices ✓. These images are then put together by the computer to form a three-dimensional image, which can be viewed from different angles ✓.

> This is an excellent answer that covers almost all the points. A comparison between the two-dimensional nature of a traditional X-ray image and the three-dimensional image from CT scanning is the only extra point that might have been included. Nevertheless there is enough here to gain full marks. 6/6

(b) (i) MRI scans do not rely on ionising radiation, unlike CT scans ✓. This makes it safer for both patient and medical staff. MRI scans give a better contrast between soft tissues than CT scans ✓.

> This excellent answer highlights two quite different advantages. 2/2

(ii) While MRI scans give a better contrast between soft tissues ✓, CT scans give better contrast with bony tissue ✓. The combination allows the doctors to relate the soft tissue to the bony tissue.

> Once more the candidate shows a clear understanding of the situation. 2/2

Question 11

(a) **When a signal is sent along a cable it is attenuated.**

   Explain what is meant by the term *attenuation*. (1 mark)

(b) **Explain the causes of attenuation in:**

   (i) **a fibre-optic cable** (1 mark)

   (ii) **a coaxial cable** (2 marks)

(c) A cable of total length 50 km has an attenuation per unit length of 5.2 dB km^{-1}. Eight repeater amplifiers are connected into the cable, each with a gain of 32 dB.

A signal of input power of 600 mW is transmitted along the cable.

Calculate:

(i)   the attenuation caused by the cable alone   (1 mark)

(ii) the total gain from the amplifiers   (1 mark)

(iii) the power of the signal after transmission.   (3 marks)

Total: 9 marks

Candidate A
(a) Attenuation is the reduction in the magnitude of a signal as it travels along a cable ✗.

🔎 The candidate has some idea of attenuation but the term magnitude is too vague here. Acceptable terms would be power, voltage, amplitude, energy. 0/1

(b) (i) Some of the signal is scattered in the glass ✓.

(ii) Heating of the cable as the current goes through ✓ ✗

🔎 The candidate has got the idea that the radiation is scattered but has not gone into detail. Nevertheless, there is just about enough for the mark. There is only one point made in the second part but it is made quite well and deserves the mark. 2/3

(c) (i) signal attenuation = 50 × 5.2 = 260 dB ✓

(ii) gain = 32 × 8 = 256 dB ✓

(iii) net loss = 260 − 256 = 4 dB ✓

$-4 = 10\ln(P_{out}/P_{in})$

$e^{-0.4} = P_{out}/600$ ✓

$P_{out} = 402 \approx 400$ mW ✗

🔎 This is fine until the last stage, where the candidate uses natural logarithms, rather than logarithms to the base 10. Notice how the candidate calculates the net loss along the cable, and correctly puts this in as a negative quantity in the equation. Enough has been done to gain a compensation mark. 4/5

Candidate B
(a) Attenuation is the reduction in the power of a signal as it travels along a cable ✓.

🔎 This is a correct description of attenuation. 1/1

(b) (i) The light/infrared radiation that carries the information is scattered by impurities in the glass ✓.

(ii) There is a current in the copper cable. Hence there is heating due to the electrical resistance of the copper ✓. Energy is also radiated away as the cable acts like an aerial with the varying current in it ✓.

📝 There are three good points here, all expressed clearly and well explained. 3/3

(c) (i) signal attenuation = 50 × 5.2 = 260 dB ✓

(ii) gain = 32 × 8 = 256 dB ✓

(iii) net gain = 256 − 260 = −4 dB ✓

$-4 = 10\lg(P_{out}/P_{in})$ ✓

$10^{-0.4} = P_{out}/600$

$P_{out} = 239 \approx 240$ mW ✓

📝 The candidate has worked through the calculation correctly, showing understanding of attenuation and gain (when measured in decibels). Unlike Candidate A, the net gain is calculated, which automatically comes up as a negative figure. 5/5